THE CARBON-14 DATING OF IRON

THE CARBON-14 DATING OF IRON

NIKOLAAS J. VAN DER MERWE

THE UNIVERSITY
OF CHICAGO PRESS
CHICAGO & LONDON

Library of Congress Catalog Card Number: 75–76206

THE UNIVERSITY OF CHICAGO PRESS, CHICAGO 60637
The University of Chicago Press, Ltd., London W.C. 1

Published 1969
Printed in the United States of America

CONTENTS

FIGURES

TABLES

PREFACE

THE PAST TWO DECADES have seen the growth of a formidable body of investigative techniques in archaeology based on the application of natural scientific principles to the study of archaeological problems. Willard F. Libby's development of the carbon-14 dating method is probably the most important milestone in this area of investigation, both in its potential scope of application and in its impact on the cultural chronologies of most areas of the world. In addition, a veritable battery of investigative techniques has been developed—a variety of relative and absolute dating methods, electronic techniques for the detection and surveying of sites, chemical and physical methods of analyzing archaeological specimens, and techniques for the intensive study of past environments. The research reported in this book adds to the growing body of physical and chemical methods used in the study of archaeological problems a technique for the study of the Iron Age. Based on the carbon-14 dating of iron artifacts through the extraction of carbon from iron alloys, this technique can be applied to the solution of cultural and metallurgical problems in the history of the Iron Age. The accuracy of this dating method, and its potential usefulness in solving historical problems, is demonstrated through the dating of a collection of iron samples of different ages and alloy types. In addition, the usefulness of dating such companion materials as iron slag (and also copper slag), when suitable iron specimens are not available, is illustrated.

An earlier version of this work was the author's doctoral dissertation, presented to the department of anthropology at Yale University. The laboratory research began in 1961 as a modest study of the feasibility of a technique for the dating of iron. Since the initial results were encouraging, a more intensive laboratory project was started in 1963 when sufficient research funds became available. In addition to the laboratory work, archaeological excavations of a number of Iron Age sites in South Africa were carried out during the summer of 1965. This field work was followed by further laboratory work, bringing the project to its present conclusion.

The blend of chemistry, physics, metallurgy, and archaeology that makes up this study reflects the cooperation so willingly extended to me by individuals from a number of diverse scientific disciplines. Professor Karl K. Turekian, of the geochemistry group in the department of geology at Yale, suggested the topic on which the laboratory research was based and guided the project through its early stages. Dr. Minze Stuiver, director of the Yale Radiocarbon Laboratory, made innumerable contributions to the design and construction of the laboratory equipment, directed the staff of the laboratory in the purification and counting of the gas samples, participated in the archaeological field project, and generally saw to it that the research proceeded smoothly. Professors William Davenport, Irving Rouse, and Kwang-chih Chang—all of the department of anthropology at Yale—were my advisors at different stages of the research; Professor D. Allan Bromley, of the department of physics at Yale, helped in outlining the program of research and in obtaining the necessary funds; Professor Cyril S. Smith, of the Massachusetts Institute of Technology, and Mr. Henry Cleere, of the Iron and Steel Institute, London, advised me on problems in the history of iron metallurgy and obtained the first iron specimens for analysis. Without the help of these individuals the project would never have gotten off the ground; their contributions are sincerely appreciated.

In the course of collecting iron specimens for the laboratory research, I received help from a great many individuals, most of whom I have never met and am therefore doubly indebted to. I wish to pay special tribute to the late Sir Ian Richmond of Oxford University, and the late Mr. William Todd of the Royal Ontario Museum, both of whom were very interested in the project but did not live to see the results. For the donation of specimens or for help in obtaining them from museum collections, I also wish to thank Dr. P. W. Bishop, Smithsonian Institution; Mr. Frank Caro, New York City; Professor J. Desmond Clark, University of California at Berkeley; Mr. Henry Cleere, The Iron and Steel Institute, London; Professor Michael D. Coe, Yale University; Dr. H. H. Coghlan, The Borough Museum, Newbury, Berkshire; Dr. Clifford Evans, Smithsonian Institution; Mr. Dudley T. Easby, Metropolitan Museum of Art; Dr. Brian M. Fagan, British Institute of History and Archaeology in East Africa; Mr. E. N. Hartley, Saugus (Massachusetts) Ironworks; Professor Hugh M. Hencken, Harvard University; Dr. I. Noël Hume, Colonial Williamsburg, Virginia; Mr. Leonard G. Johnson, Bucks County Historical Society, Doylestown, Pennsylvania; Mr. Marvin Kivett, Nebraska State Historical Society; Professor S. S. Ling, National Taiwan University; Dr. Revil Mason, University of the Witwatersrand, South Africa; Mr. Richard Muzzrole, Dr. Jacqueline Olin, Mr. John Pearce, and Dr. M. L. Peterson, all of the Smithsonian Institution; Dr. Kenneth Starr, Chicago Natural History Museum; Dr. R. F. Tylecote, University of Newcastle upon Tyne; and Mr. Peter C. Welsh, Smithsonian Institution. For the donation of specimens and permission to use them for destructive analysis, my thanks also to the trustees of the Bucks County Historical Society, Chicago Natural History Museum, Colonial Williamsburg,

Harvard Peabody Museum, Metropolitan Museum of Art, Nebraska State Historical Society, Royal Ontario Museum, Saugus Ironworks, and Smithsonian Institution.

During my fieldwork in South Africa, I received invaluable assistance from Dr. R. J. Mason, Professor Raymond Dart, Mr. Peter Beaumont, and Mr. David Seddon, all of the University of the Witwatersrand; Dr. Kearney, Mr. L. Verhoef, and Mr. J. Scholtemeyer, of Foskor; and Dr. J. Schoeman and Mr. Richard Moots of the Palabora Mining Company.

Drafts of the manuscript have been read in whole or in part by Drs. Kwang-chih Chang, R. N. Henderson, D. J. de Solla Price, Irving Rouse, and Minze Stuiver, all of Yale University; Professor Willard F. Libby, of the University of California at Los Angeles; and Mr. Henry Cleere, of the Iron and Steel Institute in London. I am extremely grateful for their many suggestions. Several drafts were typed and edited by my wife, Julia, who also participated in the field project.

The preliminary experiments on the carbon-14 dating of iron were supported by a variety of funds provided by Yale University. The laboratory project was supported by National Science Foundation grant GS–281; funds for fieldwork were provided by a Foreign Area Fellowship from the Social Science Research Council and an Augusta Hazard Archaeological Fellowship from Yale University.

THE CARBON-14 DATING OF IRON

1

INTRODUCTION

ARCHAEOLOGISTS are by now thoroughly acquainted with the carbon-14 dating method and make use of this technique as a matter of course; by the most recent count, more than eighty laboratories around the world are engaged full time in providing archaeologists, geologists, and other scientists with age measurements. The method is normally applied in dating specimens of charcoal, wood, shell, peat, or similar materials; and the results are used to reconstruct the chronologies of past climates, geologic processes, or human events. This book is a study of the application of the carbon-14 method to the dating of man-made iron and slag specimens and of the uses of such a procedure in the solution of problems in the metallurgical and cultural history of the Iron Age. It is the result of a five-year program of research in the Yale Radiocarbon Laboratory, which eventually expanded to include an archaeological field project, bringing the study to the conclusion presented here.

As frequently happens in scientific research, the idea for this new application of the carbon-14 method grew out of a fruitful thought of another scientist—in this case from a paragraph in a paper by Karl K. Turekian on the role of geochemistry in archaeology.

> There are two main problems about ancient iron and steel work to which geochemistry may be able to give some assistance. One is the type of ore used and the second is the type of fuel and reducing agent used. . . . In the case of fuels (and reducing agents) the transition between charcoal and coal is not defined in most cases (although a study of the British iron industry should prove useful here since the transformation was made within the last two hundred years). Coals have no radioactive carbon-14 while charcoals would reflect the radioactivity of the times just as it is used in carbon-14 dating. Cast irons, with about 5% dissolved carbon are particularly suitable not only to decide which fuels the ancients used but something on which a carbon-14 date would be possible if it turned out to be charcoal![1]

Turekian encouraged the author to test the feasibility of such an application of the carbon-14 dating method and suggested a direct combustion procedure as a possible means of extracting carbon from iron. (The technique which was finally used for this purpose is, in fact, an adaptation of a standard direct combustion method for the determination of carbon contents in iron alloys,[2] scaled up about 200 times.) Minze Stuiver, director of the Yale Radiocarbon Laboratory, agreed to carry out the carbon-14 measurements involved in the research and provided valuable assistance in the design and construction of the combustion equipment. With the help of these two men the first trial experiments in the carbon-14 dating of iron were carried out in 1962.

Iron Metallurgy and Carbon-14 Dating of Iron

The basic forms of manufactured iron—bloomery iron, wrought iron, cast iron, and steel—are alloys of iron and

1. Turekian 1961, p. 10.
2. Welcher 1963, pp. 654–59.

carbon, in which the distribution, concentration, and chemical form of the carbon determine most of the characteristics of the alloy in question. In bloomery iron and wrought iron, carbon occurs as bonded cementite (Fe_3C) in concentrations of about 0.2 percent or less. In cast iron, carbon may occur as free graphite, cementite, or both. Steels contain carbon in the form of cementite and in concentrations which may vary from 0.1 percent to 2 percent (hence "medium-carbon steel," "high-carbon steel," etc.); the addition of other metals (e.g., chrome) may also contribute to the characteristics of steel, but carbon remains the most important alloying material. In cast irons, carbon contents vary between 1.5 percent and 5 percent. In this book, therefore, the term "iron" is used, depending on the context, to mean specifically the chemical element iron (Fe) or to include generically one or all of the alloys of manmade iron mentioned. In the latter case, the term is used when the particular alloy or alloys can be identified from the context, or where such specific information is not required or not available.

Carbon is added to iron in the course of its production in a limited number of ways. In most techniques of manufacture, especially those which predate the Industrial Revolution, carbon in the alloy is derived directly from the fuel used in the smelting furnace or in the forge. In some modern procedures, carbon may be added at the foundry after the smelted iron has been refined. The crucible process, a technique which has been in use for many centuries, involves the heating of a mixture of iron ore and carbon in a refractory container. In all of these cases, however, carbon is added to the metal at the time of manufacture. Throughout the history of the Iron Age, this carbon additive has been, almost without exception, some form of wood (mostly charcoal) or mined coal (mostly coke).

If the carbon in an iron alloy is derived from coal (i.e., when coal or coke were used to produce it), no carbon-14 activity is detectable when the carbon is extracted. The earliest use of coal for the production of iron is well documented, however, since it occurred within the last two hundred years in most areas, and it presents no serious dating problem. When a specimen proves to have been smelted with coal, a maximum age can be assigned to it if its origin is known. On the other hand, if the carbon in an iron alloy is derived from wood (i.e., when wood or charcoal were used to produce it), a carbon-14 determination of the date of manufacture is possible. The carbon-14 activity of the extracted carbon will reflect the date of death of the wood from which it derives; since charcoal smelting (overwhelmingly the most common procedure in iron production before the Industrial Revolution) requires the use of freshly cut wood, it can be assumed that the date of manufacture is reflected by the same measurement.

Laboratory Procedure

To test the basic assumptions outlined above, and to determine the practicability of dating iron through carbon-14 measurements, an experimental (and rather crude) set of laboratory equipment was constructed for the first measurements. The first test was carried out on a specimen of modern, coke-smelted cast iron in which no carbon-14 activity is expected but which gave a measurement of about 6 percent activity, relative to a modern standard. In terms of the 3,000–year duration of the Iron Age, this meant an age error of about 400 to 500 years. Two problems which would require solution were indicated by the experiment.

1. Atmospheric carbon dioxide had leaked into the equipment and added about 6 percent of modern carbon to the sample gas. The problem lay with the combustion unit: the high temperatures involved in the extraction of carbon from iron required the use of a combustion tube of zircon, which could not be fused to the rest of the glass apparatus. The greased rubber seals which had been used to connect the combustion equipment were not leak-proof. Thus, a vacuum-sealed design was called for.

2. The usual procedure of burning small quantities of iron

in commercially available combustion boats of limited size would not suffice when 75 to 100 grams of sample material were involved. By using such equipment, more than forty man-hours were necessary to collect enough carbon dioxide for a carbon-14 measurement. Custom-made combustion equipment had to be designed.

A number of experiments with different designs finally overcame these problems, resulting in the apparatus which is described in chapter 5. The new set of equipment proved its worth in the successful dating of a series of iron specimens.

Metallurgical History of Iron

The usefulness of a dating method for iron specimens is readily obvious. The archaeologist does not always find charcoal in the desired context in his excavations, while iron specimens may be available. Furthermore, iron artifacts, unlike charcoal, are necessarily man-made and can clearly be associated with past human activities. The carbon-14 dating of iron specimens, therefore, is of significant value in obtaining more knowledge about the history of Iron Age events. That the obverse is also true became quickly apparent in the course of the research.

The suitability of an iron sample for carbon-14 dating purposes is determined by its alloy composition (specifically, its carbon content), size, and state of preservation. While size and preservation are readily observable, alloy composition can only be determined through relatively sophisticated laboratory measurements. Since archaeologists (or carbon-14 laboratories, for that matter) may not have access to the required facilities, some other predictive device is necessary to ensure a reasonable measure of success in using the dating method. A selective survey of literature on iron metallurgy and its history helped to fill this void. The treatment of this subject in chapters 2 and 3 is by no means an exhaustive discussion—it is only a guide to be used in conjunction with the dating method—but the topic has a fascination of its own. Archaeologists tend to view man's use of iron as a single quantum jump in technological evolution; the internal structure of metallurgical history in the Iron Age, however, has a clear and consistent logic. The author quickly learned that the function, origin, and cultural association of an iron artifact can provide a useful measure of its probable alloy composition—and specifically, of its carbon content. It was found, for example, that datable material of European origin is extremely scarce in contexts which predate the fifteenth century. Before that time, cast iron was not generally known in Europe; low-carbon alloys of iron were manufactured, of which relatively large samples are required for successful dating. In China, on the other hand, cast iron has been manufactured at least since the sixth century B.C. The collection of known-age specimens, which was obtained from museums for dating purposes, and the specimens which were finally dated clearly reflect these technological traditions. Samples which were suitable for carbon-14 dating were almost exclusively of Chinese origin or of the U.S. Colonial period; European museums were reluctant to part with pre–fifteenth century specimens of sufficient size. A detailed discussion of sample requirements is provided in chapter 5.

The two traditions of iron metallurgy—the "direct process" of the West and the "indirect process" of China—do not seem to be widely known outside the community of metallurgical scholars. It is hoped that by drawing attention to this problem new avenues of thought will be opened up for archaeologists who study the Iron Age. Briefly, the two traditions are as follows: In the West (before the 1380's) the basic method for the production of iron was the direct process; iron was reduced directly from its ores as bloomery iron, and steel was produced by cementation (i.e., by adding carbon to bloomery iron). In China the indirect process formed the basis of iron metallurgy; iron was reduced from its ores as cast iron, while wrought iron and steel were produced through the decarburization of cast iron. In the late fourteenth century, knowledge of the indirect process spread to Europe, where the merging of the two traditions laid the foundations of the modern iron industry. In the light of these historical events, it is obviously not enough to discuss the spread of iron technology, unless more is known about the specific techniques involved. A host of problems in the history of the Iron Age awaits solution, many of them of just this nature; the possible solution of some of these problems by the carbon-14 dating of iron is outlined in the concluding chapter of the book.

Field Applications and the Dating of Slag

As part of the problem of dating iron by the carbon-14 method, an attempt was made to apply the technique in the course of an archaeological field study. For this purpose a number of Iron Age sites in the Transvaal, South Africa, were excavated. These sites had been chosen as a result of abundant evidence in the area for the existence of a flourishing pre–colonial iron industry. Unfortunately, the conditions of preservation in the acid soil were unfavorable, and no datable iron samples were recovered. It was found, however, that the related technique of dating iron slag can serve as a valuable research tool. Reported previously by two laboratories,[3] this dating method involves the mechanical extraction of occluded charcoal nodules from slag, followed by standard carbon-14 procedures. It was also discovered that slag from the production of copper can be dated in the same manner. The results of this study are described in chapter 7.

Summary of Chapters

The material which follows consists of two distinct but related parts. Chapters 2 and 3 deal respectively with the metallurgical techniques of iron technology and the history

3. Engstrand and Östlund 1962; Östlund and Engstrand 1963; Krueger and Weeks 1965.

of these techniques. Special emphasis is placed on smelting methods, fuels, and alloys. These two chapters, which form a brief introduction to the history of iron metallurgy, are of considerable importance to the understanding of the applications of the iron dating method which is described in subsequent chapters. The theory of the method for the carbon-14 dating of iron follows in chapter 4; chapters 5 and 6 deal with the laboratory techniques used and the experimental results obtained in testing these theoretical assumptions. The dating of iron and copper slag is described in chapter 7, and the concluding chapter outlines the significance and possible uses of the carbon-14 dating of iron and slag in the solution of a number of the historical problems described in chapters 2 and 3. Those readers who require more information about the dating of iron in the laboratory will find further details about equipment and experimental results in the appendixes.

2

IRON METALLURGY

To MAKE EFFECTIVE USE of the carbon-14 dating method, the archaeologist must be thoroughly familiar with datable materials—the kinds of materials which can be dated, the quantity necessary for analysis, and factors which may contribute to inaccurate results. This is no less the case in the carbon-14 dating of iron—to use the method successfully requires at least a background knowledge of iron metallurgy and its history. Without this information it is impossible to select specimens which are suitable for dating unless metallurgical analysis is carried out. The datability of a specimen is determined by its carbon content; a reasonably accurate estimate of carbon content can be made on the basis of alloy type and weight. Alloy type, in its turn, can often be determined from a knowledge of the cultural context of the specimen.

Archaeologists, even those dealing specifically with the Iron Age, frequently have an incomplete understanding of the technical aspects of iron metallurgy. The result has been a number of misconceptions about and faulty reconstructions of Iron Age events. The results of man's ability to produce artifacts of iron are often viewed as a kind of elaboration of his skill in working nonferrous metals, which, once acquired by his Hittite forebearers, spread in the usual pattern of culture contact from one area of the world to the next, and was elaborated on in centers of technological innovation. The normal tendency has been, once evidence for iron production is documented for a given time and place, to note the fact and to look for a possible origin in adjoining areas at an earlier time level. While this view of the Iron Age is correct in many cases, it is also an oversimplification leading to errors; the development of iron metallurgy has, in fact, taken widely dissimilar courses in different parts of the globe. This lack of technical knowledge of iron metallurgy is responsible, for example, for the fact that the widely separate (and probably independent) Iron Age traditions of China and Europe are generally overlooked or accorded little significance.

The purpose of this chapter is to dispel some of these misunderstandings through a discussion of various aspects of iron metallurgy, especially on the primitive level. With this groundwork lain, the discussion of the history of iron metallurgy (chapter 3) can proceed without frequent recourse to expository asides. The reader will also be acquainted with the characteristics of different kinds of iron which the archaeologist may encounter in the course of his work. This will provide the necessary information to understand some aspects of the dating method which is described in subsequent chapters, and will point out the cases in which it can most effectively be put to use.

The discussion which follows is concentrated on the major aspects of iron metallurgy which should concern the archaeologist. Different sections deal with the sources of iron available to man; with the theory and techniques for the production of bloomery iron, wrought iron, cast iron, and steel from iron ores, as well as the properties of these materials; and with fuels, fluxes, and slags—the companion materials of iron production. The relationships between some of these materials, and the techniques by which they can be converted from one state to another, are diagrammed in figure 1.

The information in this figure is by necessity incomplete and simplified; the reader may, however, find it a useful reference for maintaining the necessary perspective of the totality of iron metallurgy as different topics are discussed.

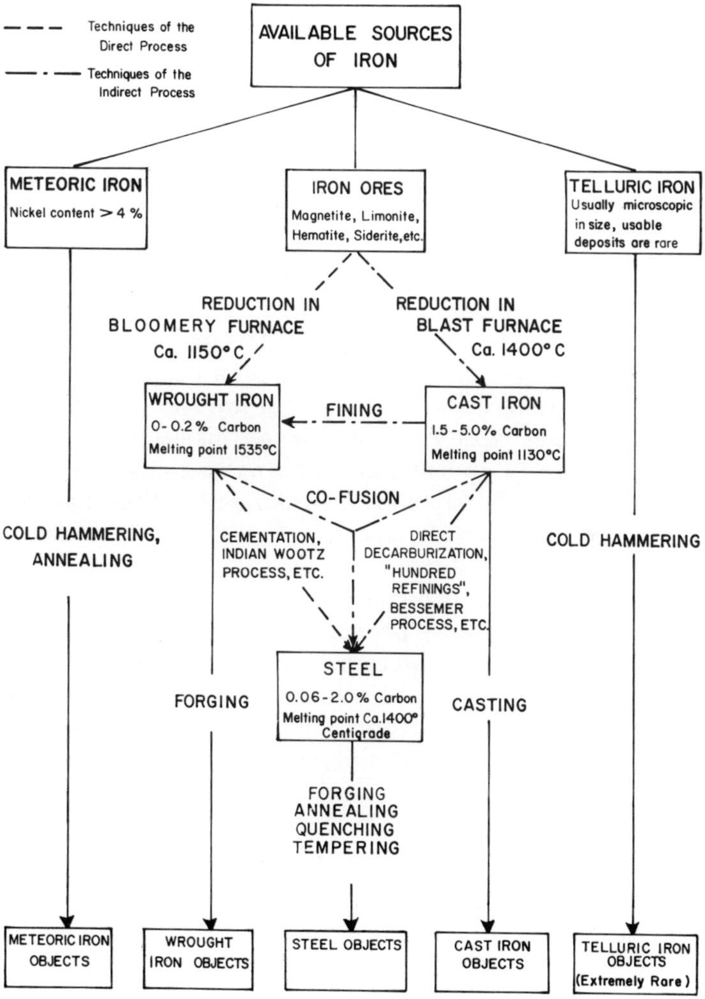

FIG. 1.—Iron metallurgy

Available Sources of Iron

Like copper, iron may occur in its native form—that is, without being chemically bonded to any other elements. In the case of native iron there are two sources available to man—terrestrial or telluric iron and meteoric iron. As a potential source for iron working, telluric iron is of relatively minor importance. It is commonly weathered out of basaltic rocks, but the resultant grains are microscopic in size and require modern technological sophistication for recovery.[1] Small, usable deposits that occur in Greenland can be cold-hammered into artifacts.

Meteoric iron is considerably more important in the history of the Iron Age; long before iron was actually smelted, artifacts were manufactured from this material, and its extra-terrestrial origin was known to some peoples. The fact that most known meteorites occur in the New World indicates that man made extensive use of this source of metal in the Old World; probability argues against any other reason for such a discontinuous distribution.[2] Artifacts of meteoric iron can generally be distinguished from other forms through chemical and structural analysis. Meteoric iron contains from 4 percent to 60 percent nickel, with most meteorites falling in the 5 to 10 percent range. While some modern steels may also have appreciable nickel contents as a result of intentional alloying, their structural characteristics are distinguishable from meteoric iron. Iron ores rarely contain more than 0.2 percent nickel.

Experiments by Zimmer[3] and Richardson[4] indicate how meteoric iron could have been utilized. When used in small pieces, it can be cold-hammered into useful shapes; larger

1. Tylecote 1962, pp. 9–13.
2. Zimmer 1916.
3. *Ibid.*
4. Richardson 1934.

pieces tend to splinter when hammered, especially if the nickel content is above 6 percent. Hot-forging presents similar problems; with the presence of more than 6 percent nickel, the material cracks along certain crystal planes. It is, therefore, doubtful that primitive people would have found meteoric iron of higher nickel content of much use. With the advent of the true Iron Age, smiths became more proficient and acquired the skills which are necessary to hot-forge most meteoric irons; where such skills were absent, low-nickel meteoric irons were probably cold-forged and annealed.

While iron objects have been manufactured from meteorites on a limited scale, the production of iron from its ores has been the more common procedure. Iron in its most useful forms occurs in the minerals hematite, limonite, magnetite, and siderite.[5] Hematite (Fe_2O_3) is a red oxide of iron, which is readily reduced and which occurs in quantity near the surface. Its hydrated form, limonite ($Fe_2O_3 \cdot H_2O$), also known as brown and yellow ocher, can be handled in the same way. Equally useful is magnetite (Fe_3O_4), a black, magnetic oxide commonly known as lodestone. The carbonate ore, siderite ($FeCO_3$), also known as chalybite, or spathic ore, has a lower iron content but presents no particular problems in smelting operations. These iron ores have a world-wide distribution; iron is second only to aluminum in abundance in the earth's crust. Different deposits of iron ore have different characteristics, determined by the presence or absence of other minerals. Most iron-bearing rocks, for example, contain appreciable amounts of siliceous materials, which must be removed during the iron smelting process; small concentrations of manganese and phosphorus in the ore may affect the smelting procedure or the characteristics of the final product. The chemical composition of the particular iron oxide involved, however, has little effect on smelting, other than the temperature required and the obvious matter of total iron yield.

5. Hodges 1964, pp. 81–82.

Bloomery Iron and the Direct Process

Bloomery iron is a soft, malleable alloy of iron and carbon with an inhomogeneous carbon content. The latter may vary from nearly zero to 0.8 percent and above in parts of the alloy; normally, the average carbon content is below 0.2 percent. The alloy is very similar to *wrought iron;* both are low-carbon alloys, containing slag stringers, but the carbon in wrought iron is homogeneously distributed as a result of the technique of its manufacture. Bloomery iron is produced through direct reduction of iron ores (hence *direct process*); wrought iron is produced through the decarburization of cast iron (hence *indirect process*).[6] Bloomery iron and wrought iron show no appreciable difference in hardness and malleability and can be distinguished only on the basis of microscopic analysis. The two are therefore often confused, with bloomery iron being referred to as wrought iron in the absence of more detailed information. The distinction, while extremely important to an understanding of the history of iron metallurgy, does not effect the carbon-14 dating of iron since there is no difference in carbon content involved. The term "wrought iron" is therefore often used in this book to designate a low-carbon alloy when it is convenient to do so; this practice is followed in figure 1.

The production of bloomery iron through the direct or bloomery process forms one of the basic elements of iron metallurgy. This primitive smelting procedure is still common in many parts of the modern world, although the twentieth century has seen a rapid decline in its use. The process involves the reduction of iron ore in a bloomery furnace; the chemical reaction involved is called smelting. For most metals, this reaction takes place above the melting point of the metal, but iron can be an exception to this rule; in the direct process, iron is reduced in its solid form.[7]

6. H. F. Cleere 1968: personal communication.
7. Tylecote 1962, pp. 183–216.

Although iron can be reduced from pure oxides at about 800°C (a temperature considerably below its melting point of 1,535°C), ores of this quality are, for practical purposes, not naturally available. Iron ores contain large amounts of gangue, or unwanted minerals, which must be removed by slagging. Slag is a combination of the oxides in the gangue (predominantly SiO_2 but also Al_2O_3, CaO, MgO, etc.) and iron oxide, usually in the form of fayalite ($2\ FeO{\cdot}SiO_2$), which can be removed from the area of ore reduction through liquation—that is, at a temperature above the melting point of the slag.[8] For most bloomery slags this process requires a temperature of 1,150°C.

The melting point of bloomery slag is well within the temperature range of a reasonably efficient pottery kiln, and certainly within that of a copper smelting furnace. On the basis of this information archaeologists are often surprised at the long time lag between the first production of copper and the advent of the Iron Age. The critical factor, which must be added here to explain this apparent anomaly, is the "carbon monoxide–carbon dioxide equilibrium ratio."[9] The reduction of iron ore by carbon fuel in a furnace depends not only on temperature but on the presence of a sufficient amount of carbon monoxide. The reduction is described by the equation

$$Fe_2O_3 + 3CO \rightarrow 2Fe + 3CO_2.$$

The proper surplus of carbon monoxide is produced through the combustion of carbon in an atmosphere, poor in oxygen, which yields two end products. Part of the carbon is fully oxidized to form carbon dioxide; the rest is partially oxidized to form carbon monoxide:

$$C + O_2 \rightarrow CO_2$$
$$2C + O_2 \rightarrow 2CO.$$

8. Cleere 1968: personal communication.
9. Read 1934*b*, p. 383.

With enough carbon monoxide present, the iron ore is reduced to iron; if the carbon monoxide supply falls below a critical level, the ore simply forms slag and no usable metal is produced. While it is relatively easy to obtain a strongly reducing atmosphere in a furnace at 800°C, it becomes progressively more difficult with increased temperatures. As the air blast is increased to speed up combustion and raise the temperature, more carbon dioxide is produced. Equilibrium between the rapid combustion which produces high temperatures and carbon dioxide on the one hand, and the incomplete combustion which produces carbon monoxide on the other, requires the presence of ore and fuel lumps of the right size, and in proper relative quantities. These critical factors could only have become known through extensive experimentation.

Early production of bloomery iron was probably carried out in simple bowl furnaces (i.e., a simple bowl in the ground, constructed of refractory clay) with the possible addition of a ring of stones. In this bowl was placed a mixture of charcoal and iron ore (the charge). The charcoal was lit, and an air blast was directed into the furnace from the side by blowing through a tube. Later elaborations saw the addition of bellows and the erection of a chimney on top of the furnace. A simple furnace for the production of bloomery iron is usually called a bloomery furnace. If a tall chimney, or shaft, is built over the furnace bowl it becomes a shaft furnace; in an induced draft furnace, the dimensions of the shaft and the porosity of the charge combine to produce a strong draft, much like a fireplace chimney. Judging from modern practices in Africa, a shaft of 5 to 10 feet must be erected to create such a draft in a furnace of any size.[10] In a forced draft furnace, the air blast is supplied by bellows, usually in pairs, and directed into the furnace through tuyeres of refractory clay.

A better understanding of the operation of a simple bloom

10. Bellamy 1904; Crawhall 1933.

furnace has been obtained through an experiment by Wynne and Tylecote.[11] Using a shallow bowl and a single tuyere, they directed an air blast into the charge over the rim of the bowl. The details of the operation are summarized below:[12]

Diameter of bowl	9 inches
Maximum temperature required	1,250°C
Air blast required	2.5 cu. ft./min.
Fe in ore	48%
Fe produced/Fe in ore	20%
Charcoal required, in lbs./lb. Fe	16 lbs.
Inside diameter of tuyere	0.5 inch
Downward angle of tuyere	45°
Total smelting time	5 hours
Ore size	½ inch
Charcoal size	¼ inch

During the experiment it was found that an ore with 5 percent lime formed a slag at 1,050°C, while a less calcareous ore slagged at 1,100°C. To let the slag run free required a temperature of 1,150°C. At the end of the smelting procedure, a lump of *cinder* (liquated slag containing lumps of charcoal) had formed in the bottom of the bowl furnace. Above the cinder, on the side of the bowl opposite the tuyere, the potential bloom (a pasty mass of reduced iron from which most of the gangue had been removed) had formed. Higher up in the furnace were reduced ore lumps and unliquated slag, which could not be used. The material in the middle layer was reheated in another furnace and hammered at a temperature above 1,050°C to squeeze out the liquid slag, thus obtaining a small iron *bloom.*

Passing from the experimental to the actual operation of a primitive bloom furnace, we find an accurate description by John Percy of its use in India during the nineteenth century.

> It [the furnace] varies in size and sometimes in shape, but the principle of its construction and the method of

11. Wynne and Tylecote 1958.
12. Tylecote 1962, p. 187.

working are always the same. In its simplest form, among the Hill Tribes of the Ghats, it is only 2 ft. high, and yields 5 or 6 lbs. of iron at a charge, while, in its more advanced state in the Deccan, it is often 4 ft. high, and produces 30 lbs. of iron at a charge. It is circular in form; its width at the bottom or across the hearth is from 10 in. to 15 in.; and at the top from 6 in. to 12 in.; and as stated above, its height ranges from 2 ft. to 4 ft. It is built entirely of clay carefully tempered. In use the lower part wears rapidly away, and is constantly repaired with a lining of fresh clay. There are two openings into the bottom of the furnace, through one of which the blast is admitted, and through the other the slag or *cinder* and the iron produced are removed. With respect to this practice Mr. Blanford informs me, that in all the furnaces which he has seen in Bengal and the Carnatic the cinder runs out at the side, and the iron is withdrawn through the same aperture as that through which the twyers pass.

The operation of smelting is conducted in the following manner:—The furnace, if a new one, is carefully dried by keeping a fire in it for some hours. Two earthen pipes or twyers about 12 in. long and 1 in. in internal diameter are placed side by side in the opening at the [rear] of the furnace, so as to project 2 in. or 3 in. into the interior of the furnace, and be 3 in. or 4 in. above the bottom, and with each of these twyers a bellows is connected. The remainder of the opening is then filled with clay, and the front opening is in a like manner stopped. The furnace is now filled nearly half full of charcoal, and upon this fire is put, after which it is filled up to the top with charcoal. The blast is then applied. The smelters attach a good deal of importance to the kindling of the fire at a proper height above the twyer, and as it extends downwards very slowly a small quantity of charcoal remains unburnt in the bottom until almost the end of the operation.

> When the charcoal sinks at the top of the furnace, alternate charges of ore and charcoal are supplied until the proper charge of ore has been introduced, after which the blast is increased as much as possible and thus maintained till the close of the operation. Cinder soon appears in the hearth, and, when it rises to the twyers, is tapped of by inserting a small bar through the fore part of the furnace. The greater part, however, of the cinder remains in the furnace and is taken out along with the iron. In from 4 to 6 hours a charge is completed, when, the front of the furnace being removed, a small mass of malleable iron, cinder, and unburnt charcoal is drawn out. If the operation has been properly conducted, the iron is sufficiently hot to be hammered at once into a tolerably sound *bloom*, with the extrusion of a thick viscid cinder; but it is sometimes too cold for this purpose, and must in that case be reheated in an open charcoal fire.
>
> As a large part of the front of the furnace is removed at the end of each successive charge to extract the iron, much time is wasted and fuel lost from the cooling of the furnace. Hence, it is generally impossible to work off from one furnace more than two or three charges daily.[13]

Except for the inclusion of slag, the iron which is produced in these primitive bloom furnaces has a high degree of purity, as a result of the low smelting temperature. The carbon content is low (below 0.2%), while the major impurities include silica, phosphorus, manganese, limestone, and aluminosilicates.[14] The metal can be forged into useful shapes through repeated heating and hammering; sufficient hammering work-hardens the material and results in a fibrous structure.

Bloomery iron is limited in its usefulness (as is wrought

13. Percy 1864, pp. 257–58 (quoted with permission of the publisher).

14. Tylecote 1962, p. 191.

iron) because it is soft; lacking the higher carbon content of steel, it does not produce durable tools which can take punishment. A typical example of its limitations is given by David Livingstone, who observed an African iron spearhead "curl up like the proboscis of a butterfly" against the head of a hippopotamus.[15] The real impetus in the Iron Age, therefore, did not come until the metallurgical processes for the hardening of bloomery iron and its conversion into steel had been invented.

Steel from Bloomery Iron

In its simplest analytical definition, steel is an alloy of iron and carbon, in which the carbon content may vary between 0.1 percent and 2.0 percent, depending on the kind of steel involved. The terms high-carbon steel, medium-carbon steel, and low-carbon steel are therefore often used, with reference to the carbon content of the alloy. The characteristics of steel depend on its crystalline structure, which is determined not only by its carbon content but also by the techniques of its manufacture. The major techniques for the production of steel include cementation, annealing, quenching, and tempering.

To convert bloomery iron to steel, carbon must be added to the alloy. This can be accomplished through the process of *cementation,* also known as *carburization.* The bloomery iron is heated to a temperature in excess of 900°C in the presence of charcoal for several hours. Carbon diffuses into the metal, altering the crystalline structure and producing a harder, but more brittle alloy.[16] In a primitive forge, where such a procedure can be carried out, the carbon penetrates the surface of the metal only. Ancient steel objects, therefore, are usually only surface-hardened, while a softer core of bloomery iron remains. The modern technique of surface-hardening ball bearings is roughly analogous to the more

15. In Butts 1963, p. 306.
16. Hodges 1964, pp. 83–84.

ancient procedure. As a result of this incomplete steeling process in ancient metallurgy, objects like sword blades were usually hammered into their final shape as bloomery iron, and then heated in charcoal to produce a carburized surface.

The hardness of carburized steel provides a distinct advantage, but its brittleness can be a problem. Further experiments with primitive steels, therefore, resulted in the discovery of the various methods of heat treatment by which the characteristics of steel can be altered. The structural characteristics of steel vary with temperature, while the properties of a piece of steel can be permanently changed by a combination of appropriate heating and cooling procedures.

When steel is produced from bloomery iron through the cementation process, stresses and strains are produced in the crystalline structure of the metal which contribute substantially to the brittleness of the product. These imperfections can be reduced by the technique of *annealing*, during which the metal is reheated and slowly cooled, often by burying the hot piece of steel in sand or ashes to slow the cooling process. The distorted crystal structure resulting from working the metal at low temperatures, which causes the brittleness, is replaced by a structure of uniform, undistorted crystals; the result is a softer, more ductile steel. The practice of *quenching* is the next logical step; it results from the fact that the varying properties of steel at different temperatures can be "fixed" into the cold structure of the material through rapid cooling. The heated metal is plunged into a bath of cold water or oil—some ancient recipes also favored the use of goat's blood or the urine of red-haired boys—to produce a harder steel. As in the case of cementation, quenching also produces brittleness, which can once more be reduced by annealing the metal. The production of a good steel, therefore, entails the application of alternate annealing and quenching procedures until the desired effect has been achieved.

If the above process is carried out under strict temperature controls, the desired effect can be easily obtained and duplicated. Such controls can be applied by observing the color of the steel as it is heated, and quenching it when the desired color (i.e., temperature) has been reached. This process is known as *tempering*, and it is carried out in a range of temperatures from 100°C to 650°C. The color changes in steel fall within narrow temperature ranges, thus providing accurate controls of the properties of the final product. Blue steel, suitable for the manufacture of chisels and saws, is the result of tempering in the 290° to 330°C range; purple steel, suitable for swords and knives, is produced between 270°C and 290°.[17] As can be expected from the logically progressive relationships between cementation, annealing, quenching, and tempering, these techniques form an evolutionary sequence with several centuries separating the appearance of each in iron metallurgy.

The repertoire of metallurgical techniques in the production of steel is endless, but two further procedures deserve mention here, since they result from a direct application of the cementation process. *Pattern-welding* is an ancient European process, known for its use in the manufacture of Merovingian and Viking swords. These sword blades have a pleasing, wavy-lined pattern along the central portion of the blade as a result of alternating areas of high and low carbon content. This result is obtained by welding together strips of bloomery iron and steel, or strips of bloomery iron carburized on only one side, to give a rod of laminar structure.[18] The rods are then twisted, folded, and forged into a strip, which eventually becomes the spine of the sword blade. When the completed blade is polished the difference in surface texture of bloomery iron and steel creates an impression of light and dark bands; this effect can be increased through etching with acid, or by allowing the blade to tarnish.[19]

17. Hodges 1964, p. 84.
18. Smith 1960, pp. 3–4.
19. Hodges 1964, p. 88.

Pattern-welding is thought to be an attempt at copying the process of *damascening*, the technique used in producing the famous Damascus sword blades. The surface of a Damascus blade looks like watered silk, an effect which is once more the result of etching an object with areas of varying carbon concentration. The technique of manufacture, however, differs from pattern-welding in that the watered structure is obtained before the blade is forged and that the blade is manufactured entirely of steel of varying carbon concentration. Representative pattern-welded swords have an average carbon content in the order of 0.1 percent; Damascus blades are made of high-carbon steel with a carbon content of about 1.5 to 2.0 percent. Although damascening shows many local variations, the Indian process of manufacturing *wootz* is representative of the techniques which produce the characteristic watered-silk pattern. The basic material is an iron bloom, which is broken down and packed together with wood and leaves in clay crucibles. The crucibles are sealed with clay, dried, and then fired at a high temperature and allowed to cool slowly. Ingots of brittle, high-carbon steel, in which the wavy pattern is already present, are produced. The effect is heightened by alternately heating the steel in a current of air and forging it, thus removing some of the carbon. The result is a steel composed of cementite (iron carbide, Fe_3C, in which the carbon has combined with the metal) distributed in pearlite (iron with free, occluded carbon, and in lower concentrations), which produces the Damascus pattern when the metal is forged into a blade.[20] The pattern is made visible through etching with a solution of mineral sulphate.[21]

Since the industrial revolution, metallurgical techniques have proliferated and the number of steel alloys for different purposes has increased sharply. The above discussion, however, includes the major topics which concern the archaeologist.

20. Hodges 1964, p. 89.
21. Smith 1960, p. 16.

Cast Iron and the Indirect Process

Cast iron is a brittle alloy of iron with a carbon content ranging between 1.5 percent and 5 percent, which occurs in two forms—white iron and grey iron. White iron is the more brittle and less desirable of the two, containing carbon in the form of the brittle material, cementite (Fe_3C). Grey iron, on the other hand, contains areas of free carbon in the form of graphite, while the surrounding areas consist of fairly pure wrought iron or steel. The difference between grey iron and white iron is determined by their relative silicon content; high-silicon grey iron is produced at high temperatures, while low-silicon white iron is produced at relatively lower temperatures.[22]

The best known attribute of cast iron is its melting point (*ca.* 1,150°C), which is lower than that of steel (*ca.* 1,400°C) or wrought iron (1,535°C) and allows it to be cast in the molten state. While the discovery of iron casting added a significant new dimension to the existing techniques of iron metallurgy, the real significance of cast iron involves the economics of iron production. The metal yield from the ore is greater when cast iron is manufactured, while steel can be mass-produced through oxidation of the carbon in cast iron—a procedure which is considerably less laborious than cementation. Wrought iron can be manufactured in the same way by carrying the oxidation a step further; this is known as the indirect process.

Although cast iron may have a melting point as low as 1,130°C, depending on its carbon content, this temperature must not be confused with the much higher temperature at which it is produced. A number of variables combine to make a temperature as high as 1,400°C a requirement for the production of cast iron. The high carbon content of cast iron can only be obtained through a long period of contact between ore and fuel at temperatures of more than 1,225°C.

22. Tylecote 1962, pp. 302–3.

At these temperatures, more of the iron oxide is likely to be converted to slag. This tendency can be corrected by the addition of flux, which replaces the iron in the slag but which may raise the free-running temperature of the latter. In addition, temperatures of well over 1,200°C are required to absorb lumps of, for example, limestone flux in the slag. A further complication in reducing iron ores at such high temperatures involves the carbon monoxide–carbon dioxide ratio. The problem of maintaining a high temperature as well as a reducing atmosphere, already difficult in the bloom furnace, becomes critical when cast iron is manufactured.

The solution of the problems described here requires a sophistication in furnace construction and operation which took the Europeans more than 3,000 years to develop; only in China was the appropriate furnace in use before the Christian era. Cast iron is manufactured in a *blast furnace*, which represents a refinement in design of the shaft furnace. In the blast furnace the ratio of height to diameter is considerably increased to allow for longer contact of the ore with charcoal and reducing gases. Ancient blast furnaces were generally taller than six feet, while a height of more than fifteen feet is not uncommon. In addition, a continuous air blast of considerable force must be applied to attain the desired furnace temperature. In the European design waterpower was generally used to move the bellows and produce the blast. The Chinese, with previous experience in the use of fire-pistons, used a double acting (push-pull) piston to achieve the same result.

The chemical reaction in a blast furnace is the same as in a bloom furnace. Iron ore in the upper part of the furnace is reduced by carbon monoxide coming up the shaft. Prolonged contact between the iron and the fuel in the furnace results in the absorption of carbon in the iron. As more carbon is absorbed, the melting point of the iron is lowered, and a fluid metal results. The molten metal sinks to the bottom and can be tapped off and cast into useful forms at the foot of the furnace. While this process uses large amounts of fuel—a fact which eventually necessitated the use of coke as a smelting agent—the amount of iron recovered from the ore is much higher. Typical figures from seventeenth-century England show blast furnace yields of 92 pounds of iron per one hundred pounds of iron in the ore, as compared to a contemporaneous bloomery furnace yield of 55 percent.[23]

While cast iron can be cast into shapes which could not readily be produced by other means, it is brittle and thus unsuitable for a large variety of purposes. Malleable wrought iron, however, can be produced in quantity by heating cast iron in an oxidizing atmosphere. This process, known as *fining*, can be carried out in a simple bloom furnace, where the required temperature can easily be attained if an oxidizing atmosphere is desired. To produce a wrought iron bloom, a bar of cast iron is held over the hot charcoal and allowed to melt and drip from the end. The carbon in the drops of iron is rapidly oxidized as they fall; the melting point of the iron is raised as a result of decreased carbon content and a solid bloom collects in the bottom of the furnace.[24] Steel can be produced by directing an oxidizing air blast over molten cast iron in a clay crucible. The principle is the same as for the finery, except that the degree of decarburization can be controlled to a certain extent to produce steel of different carbon concentrations. The well-known *Bessemer process* is merely a refinement of this procedure: cast iron is fined in a "Bessemer converter" by blowing air through the molten metal. Steel can also be produced by the *co-fusion* of cast iron and wrought iron; heated wrought iron is immersed in molten cast iron to produce a steel with a carbon content which is an average of the carbon contents of the original materials.

Fuel, Flux, and Slag

Iron Age archaeologists are naturally most concerned with the characteristics of the iron objects they encounter in excavations and the nature of the furnaces in which they

23. Tylecote 1962, p. 302; Schubert 1957, p. 245.
24. Tylecote 1962, p. 306.

were produced; knowledge of the fuel and flux involved in the smelting operation, however, as well as the slag which is produced, is necessary for a complete reconstruction of the metallurgical process used. The additional fact that these materials are preserved in greater quantities than iron in archaeological deposits argues further for a wider understanding of their properties and technological implications.

Until quite recently the common fuel for the smelting of iron in most parts of the world was charcoal. The charcoal was prepared in quantity from wood with the sap still in it, with iron smelting eventually becoming such an overwhelming consumer that large areas of Europe were deforested.[25] The bloomery process requires about sixteen pounds of charcoal to produce one pound of iron and about four pounds of slag; in the blast furnace, the consumption of charcoal is even higher.

While charcoal does not provide the high temperatures which can be obtained with mined coal, it has specific advantages which favor its use as a smelting fuel. Depending on the kind of wood used in the preparation of charcoal, as much as 20 percent of the fuel may be ash, of which the alkali content has been shown to be as high as 22 percent.[26] Since all the ash is absorbed in the slag during smelting, the melting point of the slag will be lowered by as much as 100°C if the alkali content of the slag is 3 percent. A further advantage of charcoal as a smelting fuel is its low sulfur content, since high sulfur contents seriously effect the quality of iron. Charcoal-smelted irons usually contain less than 0.05 percent sulfur, with an occasional concentration of 0.1 percent; coal-smelted iron will show a sulfur content of more than 0.1 percent, while measurements in the 0.4 percent range are not uncommon.[27] Coal did not become a reliable smelting agent before the advantages of coking it were discovered.

The use of coal as a smelting agent has been well docu-

25. Schubert 1957, p. 219.
26. Tylecote 1962, p. 190.
27. *Ibid.*, p. 191.

mented in some areas and its history will be discussed in more detail in chapter 3. In other areas, however, knowledge of the use of coal is vague, at best. The systematic use of coal in iron smelting in Europe is known to have been initiated in England in the early eighteenth century with the discovery of the use of coke, but its occasional use in uncoked form in Roman times is often hinted at. In China, coal may have been used for this purpose before the Christian era. Coal-smelted iron, of course, can usually be identified through a determination of its sulfur content. A more reliable method, involving a carbon-14 measurement of the carbon in the iron, will be discussed at length in later chapters.

The nature of slag, the by-product of smelting operations, is intimately tied to the use of flux. It has been mentioned above that fluxes may lower the melting point of slag and thus facilitate the reduction of iron ore at lower temperatures; its primary purpose, however, is to replace iron oxide in the slag, especially in the blast furnace, giving rise to a greater yield of iron from the ore. The use of flux for the purpose of lowering the melting point of slag has generally been overrated by investigators.[28] While a moderate amount of silica or limestone flux will lower the melting point, a point of diminishing returns is soon reached, and thereafter a significant increase in melting point temperature is brought about. In addition, large pieces of flux (e.g., limestone) cannot be absorbed by slag at temperature below 1,200°C. The identification of the use of flux at primitive smelting sites, therefore, had best be viewed with some doubt. In blast furnace operations, of course, fluxing is intended to replace iron in the slag at high temperatures, and its use can almost be taken for granted.

Although slag is preserved in large quantities at smelting sites, it has not been studied in great detail. It can generally be recognized that three kinds of slag are produced in a bloomery furnace at different stages of the furnace operation, but these are only now receiving the proper experimental at-

28. *Ibid.*, p. 187.

tention in a study by Tylecote.[29] Two major types of slag can be mentioned at this point. True *tap slag* is produced in a high-temperature furnace and tapped off in the liquid state in quantity. It occurs in large lumps and has a glossy, rippled surface. Slag of this kind is not produced in the more primitive types of bloomery furnace, of which cinder is more characteristic. Cinder is formed in the bottom of these furnaces, has an uneven, angular surface, and contains lumps of unburnt charcoal. The two types are chemically identical (fayalite), but are easily identifiable on morphological grounds.

The discussion of iron metallurgy in this chapter should provide the reader with the necessary technical background for the discussion of the history of iron which follows. The characteristic compositions of different iron alloys and slags, coupled with their distribution in time and space, will serve as points of reference in the selection of specimens which can be dated by the radiocarbon method.

29. Cleere 1968: personal communication.

3

THE PREHISTORY AND HISTORY OF IRON METALLURGY

THE HISTORY of man's use of smelted iron is a story of slow and sporadic technological development. It is characterized by an early discovery (*ca.* 2700 B.C.) which failed to make any appreciable impact; some fifteen centuries later, iron finally achieved prominence; the knowledge of its production spread rapidly, and two quite distinct metallurgical traditions developed in the East and the West.

Before the knowledge of iron smelting was acquired, man was able to use meteoric iron. Skills developed in cutting and grinding stone, common since Neolithic times, were sufficient to fashion meteoric iron objects. The knowledge of reducing iron from its ores, however, was not acquired until the third millennium B.C. The resultant metal was lacking in quality, and only isolated occurrences in Anatolia, Mesopotamia, and adjoining areas, have been recorded. Bronze, then only in its early stage of cultural significance, proved to be both cheaper and more durable than the early forms of iron for the fashioning of cutting edges. The influence of iron as an important material of manufacture was not felt until the development, by Hittite subjects, of the basic techniques of steel production. After an initial period of development, during the course of some five centuries prior to

1200 B.C., iron spread rapidly. By 500 B.C., iron was in use in most of Europe, in the Far East, and in Africa as far south as Nubia and Nigeria.

As the technique of iron smelting became more widely known, new metallurgical procedures were added. In the Mediterranean area, techniques for the manufacture and improvement of steel developed rapidly. By the beginning of the Christian era, the techniques of carburization, annealing, quenching, and tempering were widely known and the use of the direct process had become firmly established. In China, a different metallurgical tradition arose; as soon as the utility of iron became known, cast iron was manufactured. The process of steel production through decarburization was quickly developed and became the hallmark of iron metallurgy in the East. In Europe, the direct process held full sway until the fourteenth century, when the introduction of cast iron and the indirect process laid the foundation of the modern iron industry.

Meteoric Iron In History

The prehistoric use of meteoric iron in the manufacture of tools and decorative objects probably had little or nothing to do with the development of iron-smelting techniques. Although a knowledge of annealing helps to improve the quality of a meteoric iron object, the skills of stone cutting and grinding and of cold-hammering are sufficient to obtain a reasonably good product. Peoples with a relatively low level of technological development are known to have made use of meteoric iron; the Eskimos of Greenland were observed in this pursuit by polar explorers in recent times.[1] It is therefore evident that the working of meteoric iron contributed but little to the tradition of iron metallurgy. Nevertheless, man's first encounter with the properties of iron came through his use of meteorites, and its earliest use deserves some discussion here.

1. Zimmer 1916.

The list of early meteoric iron artifacts in the archaeological record is fragmentary and brief. This is due, in part, to the fact that available sources of meteoric iron are extremely limited and that a correspondingly small number of artifacts have been manufactured from this material. Equally important is the fact that a chemical determination of the nickel content of a piece of iron or a metallographic analysis of its structure is required to identify its meteoric origin; it is significant to note that meteoric iron objects have generally been identified only when large-scale archaeological projects, in which experts from many disciplines participated, were involved, or in cases where iron artifacts occur unexpectedly early in the archaeological record.

From a list of early meteoric iron objects, compiled by Coghlan,[2] a number of examples deserve mention here. The earliest known occurrence comes from Gerzah in Egypt, where Wainwright[3] discovered a number of iron beads. These beads were dated, according to the Petrie system, to S.D. 60–63 (*ca.* 3500 B.C.), and have a nickel content of 7.5 percent, clearly in the meteoric iron range. In Mesopotamia, Woolley recovered iron fragments with a 10.9 percent nickel content[4] from the Royal Tombs of Ur (*ca.* 2500 B.C.). At the Anatolian site of Alaca Hüyük,[5] two iron specimens with 5.08 percent and 4.3 percent nickel, respectively, have been identified in the Early Bronze Age II levels (*ca.* 2600–2300 B.C.).

Some of these early specimens, notably in the case of Alaca Hüyük, were contemporary with smelted iron objects in the same deposits. It is reasonable to assume, therefore, that many objects of meteoric iron have gone unnoticed for lack of chemical or metallographic analysis. Knowledge of the use of this material is likely to remain confined largely to times and places where iron objects occur in an unexpected context.

2. Coghlan 1956, pp. 32–36.
3. Wainwright 1936, p. 7.
4. Peake 1933, p. 641; Desch 1928.
5. Kosay 1944, p. 189.

THE INVENTION OF IRON SMELTING: THE BLOOMERY PROCESS

While the use of meteoric iron appears to have made a minimal contribution to the development of iron metallurgy, the traditions of nonferrous metallurgy played a decisive role. Once the techniques of producing gold, silver, copper, and metals had become known to man, the discovery of iron as a useful material was a logical development. It is no accident that iron was first smelted in those regions—the Near East and Anatolia—where copper and bronze metallurgy had its strongest early base.

Some attempts have been made, by inductive reasoning, to reconstruct the procedures by which iron was first smelted. The simplest of these reconstructions involve the production of gold from gold-bearing sand.[6] The ancient Egyptians melted gold from Nubian desert sands, which also contain quantities of magnetite. Under the proper set of conditions, reduced iron would form on top of the melted gold in the crucible, and under a layer of iron slag. This would take place if a reducing atmosphere were used accidentally and if the ratio of magnetite to sand were in the order of 2:1—a situation which could result if a flotation or washing process were used to purify the sand.[7] The iron so produced would, of course, be solid and may well have been discarded. The terms for meteoric iron and smelted iron in Ancient Egyptian clearly indicate, however, that the relationship between the two was known; knowledge of meteoric iron may have enabled the gold smelters to recognize smelted iron.

Another hypothetical reconstruction can be made by considering how iron may have been smelted accidentally in a copper furnace.[8] When copper sulfide ores are roasted before

6. Forbes 1950, pp. 403–4.
7. Cleere 1968: personal communication.
8. Tylecote 1962, p. 184.

smelting, they are converted to a reddish oxide, not unlike hematite in appearance. If the smelter used hematite in his furnace instead of copper ore, and under reducing conditions, he would obtain useless, molten iron slag in the furnace bottom instead of molten copper. If, however, he paid attention to the reduced lumps of iron immediately above the slag, he would have found these to be malleable above 1,000°C. While this last phase of the argument may be difficult to accept, it seems reasonable to argue that the idea of producing metals from mineral ores would have prompted experimentation with a variety of ores. In fact, it may not be necessary to postulate the accidental charging of a furnace with iron ore; deliberate experiments with different ores may have taken place. Familiarity with the properties of meteoric iron may have facilitated the recognition of iron as a usable material after it had been produced, accidentally or intentionally, a number of times. The fact that the earliest iron was produced in the early phases of the Bronze Age, and in those areas with the most accomplished metallurgical industries, strengthens the viewpoint of deliberate experimentation. It should also be considered that iron was regarded as a precious metal for many centuries after its discovery; the earlier economic success of the production of gold and silver probably provided considerable incentive toward the discovery of metals which could confer similar monetary rewards on the successful smelter. While the exact procedure by which iron was first smelted can thus only be guessed at, we do know what new techniques were required to produce a usable object from these early products of the bloomery process. "The discovery of man-made iron . . . had not awaited the evolution of a basically new smelting process; it was almost entirely the outcome of hammering a hot, spongy aggregate of metal, slag and dirt."[9] Thus the arts of the blacksmith were born, beginning a long period of technological evolution which was eventually to give rise to the Iron Age proper.

9. Aitchison 1960, p. 101.

Our knowledge of these early forays in iron metallurgy is fragmentary and is based on a number of isolated finds, which occurred under unusual conditions favoring preservation. From a detailed list of early iron objects provided by Coghlan, a few specimens from the third millennium B.C. can be mentioned here.[10] (These specimens can all be inferred to be bloomery iron). At Tell Chagar Bazar, in Northern Syria, Mallowan excavated a fragment of iron which dates from about 2700 B.C.;[11] at a higher level of the site, dating to *circa* 2500 B.C., two more fragments occurred.[12] Frankfort's excavations at Tell Asmar, Iraq, yielded an iron knife blade in a bronze hilt dating to the end of the Early Dynastic period of Sumer (*ca.* 2450–2340 B.C.).[13] Another important specimen, a dagger with an iron blade and a gold-sheathed handle, comes from the Royal Tombs at Alaca Hüyük, Anatolia, and dates from *circa* 2600–2300 B.C.[14] These early iron objects would have been too soft to produce durable cutting edges and were probably of ritual rather than utilitarian importance. In addition, the price of iron during the early stages of the Hittite confederacy (early second millennium B.C.) is known to have been five times that of gold and forty times that of silver,[15] and it must have been even more expensive during the third millennium B.C. At such prices, iron objects are likely to have been traded as marks of status among the royalty of ancient Near Eastern kingdoms, thus achieving a distribution which is much wider than the actual areas of manufacture. To complicate the picture further ". . . objects of third and early second-millennium iron so far recorded are far too scanty to permit of a distribution map which would be of any real value in determining the focus of the first iron working. We can only indicate its probable origin

10. Coghlan 1956, pp. 61–67.
11. Mallowan 1936, p. 26.
12. Mallowan 1937, p. 98.
13. Frankfort 1950.
14. Kosay 1951, p. 167.
15. Aitchison 1960, p. 102.

as somewhere to the north, or north-east, of the Fertile Crescent."[16]

Iron disappears from the archaeological record following the Early Bronze Age in the Near East and Anatolia. Whether this disappearance is more apparent than real is a subject for further archaeological study. We do know, however, that when iron once again appeared on the scene at the time of the New Hittite Empire the tradition of iron metallurgy had undergone a major shift as a result of the production of steel.

Prehistory and History of the Direct Process

A major impetus of the Iron Age proper was the discovery of cementation, the technique by which steel can be produced from bloomery or wrought iron and which is generally associated with the bloomery process. The discovery of this technique is usually attributed to the Chalybes, subjects of the Hittites, and dates to about 1500–1400 B.C. (The assumption that steel was an invention of the Chalybes was made by the Greeks, hence the Greek word for steel, χάλυψ.) The Hittites are thought to have maintained a strict monopoly on the manufacture of the new alloy, enabling them to keep prices at an artifically high level. This viewpoint is based on interpretations of a letter from the Hittite king, Hattusilis III (1281–1260 B.C.), to an unknown correspondent and is the subject of some dispute. Nevertheless, the occurrence of iron of any kind outside the Hittite empire is extremely limited at this time. A few pieces were obtained by Egyptian rulers; these items were regarded as of sufficient value to be included in the tomb treasures of Tutankhamen. By 1400 B.C., iron was also being traded to the Assyrians, who applied their metallurgical skills to produce even better steel than the Hittites.[17] Whether the Hittite restrictions on the iron trade were a matter of calculation or of inefficient production

16. Coghlan 1956, p. 68.
17. *Ibid.*, pp. 62–69.

remains conjectural; the spread of iron-working techniques, however, did not commence until the fall of the Hittites. After being weakened by the power of Egypt and Assyria, the Hittite empire succumbed to invasions of Thraco-Phrygian peoples from southeastern Europe (*ca.* 1230 B.C.), and the spread of the Iron Age began. The archaeological record does not provide sufficient information to determine the extent of the spread of all the available techniques of iron production (specifically cementation). The available information will be taken up in detail in a subsequent section of this chapter; in the present discussion it will be assumed (unless otherwise specified) that the techniques of the direct process were disseminated simultaneously, while keeping in mind that knowledge of the cementation process may have lagged behind knowledge of the bloomery process in peripheral areas.

During the period 1200–900 B.C., use of iron for such utilitarian objects as weapons and, in some cases, agricultural tools, spread rapidly to Syria, Palestine, Persia, Mesopotamia, and Egypt—on the one hand—and Crete, Cyprus, and Greece on the other.[18] This rapid spread was facilitated by the large-scale movements of people around the Mediterranean at the time; the tradition of itinerant smiths, already established during the Bronze Age, also aided in transmitting the knowledge of iron-working to new areas.[19] By 1000 B.C., the Phoenicians had introduced iron to their coastal settlements in Northwest Africa;[20] Etruscan immigrants from the Eastern Mediterranean carried knowledge of the new metal to Tuscany by about 800 B.C.;[21] Greek colonists served the same function in South Italy. In Austria, where it probably arrived from Asia Minor via the Danube route,[22] iron contributed significantly to the material culture

18. *Ibid.*
19. Aitchison 1960, p. 117.
20. Gabel 1965, p. 70.
21. J. G. D. Clark 1952, p. 200.
22. Cleere 1968: personal communication.

of the Halstatt peoples (from about 650 B.C. on), who spread its use in rapid succession to the Low Countries, France, and Iberia to the west, and the Middle Danube basin and Yugoslavia to the east.[23] The Halstatt peoples, however, continued to rely on bronze as a major source of metal; it was only in the subsequent La Tène cultures of Europe that iron became predominant. Originating around 425 B.C. among the Celtic-speaking peoples in the area between the Upper Danube and the Marne, the iron-working techniques of the La Tène cultures spread east into the Danube Valley, south into Cisalpine Gaul, and west and north to France, the Low Countries, and the British Isles,[24] where the tradition continued until the period of Roman expansion.

South of the Mediterranean, iron found fertile ground in several African centers. The early use of iron in Egypt (*ca.* 1200 B.C.) and Northwest Africa (*ca.* 1000 B.C.) has already been mentioned; information about the spread of iron to the rest of Africa is subject to a number of problems regarding routes of transmission, while detailed studies of the specific metallurgical techniques involved are nonexistant. It is, however, possible to obtain from archaeological sources an outline of the transmission of iron-working knowledge through the continent. In this regard, the Napatan-Meroitic civilization of Nubia, which flourished from about 800 B.C. to A.D. 350, has often been mentioned as the center from which iron was diffused to Sub-Saharan Africa. This assumption appears to be based on the presumed influence of the extensive iron industry which is known to have developed at Meroë. However, this industry did not develop early enough to warrant such an assumption. Small amounts of iron, apparently trade items, appear in Nubian graves by 400 B.C.; the development of a local iron industry did not take place until about 100 B.C.[25] This industry continued until the sack of Meroë (*ca.* A.D. 350) by the Ethiopian Kingdom of

23. G. Clark 1961, p. 170.
24. *Ibid.*, pp. 171–72.
25. Wainwright 1936.

Aksum, where iron had been introduced from across the Red Sea and where a similarly strong industry had developed. In view of the early occurrence of iron in West and East Africa, however, neither of these kingdoms could probably have contributed significantly to the diffusion of iron to the rest of Africa. It seems more reasonable to postulate a spread of iron-working knowledge to West Africa from Northwest Africa, where it had been available since 1000 B.C. Carbon-14 dates indicate that iron may have been produced in Ghana as early as 1000 B.C. and that it was known in the Nok culture of Nigeria by about 500 B.C.[26] From here it probably spread to Central, East, and South Africa. Once again, the nature of this transmission is unclear, although the archaeological record indicates that the appearance of iron in the subcontinent is closely linked with the earliest occurrences of pottery and agriculture. It can probably be assumed that the diffusion of this cultural complex is closely linked with the migrations of Bantu-speaking peoples from a center in Eastern Nigeria. In general, knowledge of prehistoric Iron Age cultures of Sub-Saharan Africa is incomplete; let it suffice to say that iron is known to occur in South-Central Africa by about A.D. 1,[27] in Zambia during the first century A.D.,[28] in Rhodesia before A.D. 300,[29] and in the Transvaal, South Africa, by about A.D. 800.[30] The metallurgical techniques involved in this transmission of iron-working knowledge are not known; it can be assumed that both bloomery iron and steel were manufactured in most areas, although modern ethnographic examples indicate that the cementation process was certainly not known to all Iron Age peoples in Africa.

Spreading east from the Mediterranean, iron was diffused throughout most of Asia before the Christian era. By 1100

26. Stuiver and van der Merwe 1968.
27. Gabel 1965, p. 70.
28. Fagan 1965, p. 52.
29. *Ibid.*, p. 58.
30. Stuiver and van der Merwe 1968.

B.C. it was in use in Persia, from where it spread to Pakistan and India. The date of the arrival of iron in India is still a matter of some dispute; until recently, iron was assumed to have reached Northern India around 500 B.C., where it appears at the sites of Taxila, Histināpura, and Ahichchhatrā in association with the distinctive "Northern Black Polished" pottery type.[31] Recent excavations at Atranjikhera, in Uttar Pradesh, however, have uncovered iron artifacts in association with "Painted Grey" pottery, from an earlier period of the Ganges civilization, and have been dated between 1100 and 1000 B.C.[32] Further archaeological work will be necessary to gauge the impact of iron-working knowledge on Northern India, especially regarding the forces which contributed towards the urbanization of the peoples of that area. In the southern part of India, at least, especially in the Deccan, iron seems to have stimulated a veritable "revolution" of this kind.[33]

The transmission of iron-working knowledge to China, if it took place at all, is a problem which remains unsolved. The possibility exists that iron was carried to China by the nomadic tribes of the Eurasian steppes. During the latter half of the first millennium B.C. the Sarmatians, a tribe closely related to the Scythians, are known to have occupied the region adjoining Kansu, in Northwest China.[34] The Sarmatians relied primarily on bronze as a source of metal, although using iron to a limited extent. Their penetration to Northwest China is marked by the appearance of their distinctive "Animal Art Style" in Mongolia and Ordos, where it is dated to *circa* 500 B.C.—and possibly earlier. Since iron appears in China during the sixth century B.C., and perhaps earlier,[35] it is doubtful whether the Sarmatians did, in fact,

31. Wheeler 1959, pp. 24, 131; 1966, p. 113.
32. Wheeler 1966, p. 113; Agrawal, Kusumgar, and Lal 1965.
33. Wheeler 1966, p. 149.
34. Phillips 1965, p. 30.
35. Kwang-chih Chang 1963, pp. 195–202.

carry the knowledge of iron to China. If they did, it was at best a case of stimulus diffusion, since the Chinese did not adopt the direct process, which had been the exclusive method for the production of iron until that time. Instead, cast iron was apparently manufactured in China from this early date on, and the techniques of the indirect process were evolved.[36]

In the New World, iron cannot be said to have achieved any widespread use before Colonial times. Small quantities of trade iron, however, penetrated to Northern Alaska by way of Siberia. Iron has been found in an early context in a site of the Ipiutak culture at Point Hope, Alaska;[37] on the other side of the Bering Strait, iron occurs in an Old Bering Sea site at Uelen on the Chukchi coast.[38] Both of these cultures have been dated to about A.D. 300.[39] Iron was not manufactured in the New World, however, until Viking colonists introduced it to Newfoundland around A.D. 1000.[40]

Evolution of the Western Tradition of Iron Metallury

The prehistory and history of iron metallurgy in the West before the late fourteenth century revolves around the production of bloomery iron through the direct process and the production of steel through cementation. The invention of the steeling process in the Hittite empire provided a major impetus in iron metallurgy, but it did not solve all the problems in the production of quality steel. In fact, it only provided the first step in the evolution of the blacksmith's art.

Detailed knowledge of the development of smithing is limited, but it is possible to obtain an outline of this evolu-

36. Needham 1958, p. 3.
37. Larsen and Rainey 1948, p. 83.
38. Levin and Sergeyev 1964.
39. Rainey and Ralph 1959.
40. Ingstad 1964.

tionary sequence in a number of areas. Studies of the metallography of iron from Egypt and adjoining areas,[41] for example, have produced sufficient analytical data to allow for a reconstruction of the standards of technological achievements at various times when the direct process was in use.

In Egypt, steel tools of uneven carbon content (i.e., with carburized areas confined to the cutting edges or the surface) were being produced by 1200 B.C. By 900 B.C., the Egyptians had learned of the advantages of quenching and were treating cutting edges in this manner. A certain degree of self-tempering, a result of heat conduction from the unquenched center of the object, is evident at this time. It is only during the Roman period that intentional tempering, albeit of a poorly executed variety lacking temperature control, can be noticed in Egyptian artifacts. Since the carburization of bloomery iron was literally only surface deep, the technique of producing laminated structures from thin sheets was developed. As early as the eighth century B.C., this technique can be found in Syria, while seventh– and sixth–century B.C. artifacts from Syria and Palestine show as many as fifty laminations in a single blade. The brazing of iron with bronze alloys also occurs at this time.

In contrast with the early knowledge of hardening carburized iron in Egypt and the Near East, the Iron Age cultures to the west were relatively unsophisticated at this time. Analysis of Halstatt swords (*ca.* 500 B.C.) as well as La Tène material from Switzerland and Italy (as late as A.D. 50) show a lack of any hardening of the basic carburized metal. Even so, the Halstatt material is brazed, and generally indicates the workmanship of highly skilled smiths.

The period of Roman expansion played a decisive role in the transmission of metallurgical techniques in Europe. The Romans were familiar with the use of carburization, quenching, and a crude form of tempering, and carried this knowledge to all areas under Roman influence. In addition, the

41. Carpenter and Robertson 1930; Coghlan 1956, pp. 134–65.

large-scale use of iron by the Roman legionary forces stimulated the development of local iron industries of some size, as attested by the excavation of a seven-ton cache of nails at a Roman fort in Scotland.[42] The collapse of the Roman Empire in the fifth century had an adverse effect on iron metallurgy in many parts of Western Europe for the next two hundred years, but it continued in strength in the Byzantine Empire (Greece, Asia Minor, the Near East, Egypt, and Nubia).[43] Further east, the iron metallurgy of India was also growing apace, as seen in the manufacture of the famous pillars of Delhi and Dhara. Erected around A.D. 300, these pillars are constructed of solid bloomery iron and weigh more than six tons apiece. The manufacture of Indian *wootz* steel for so-called Damascus blades, already known before the Christian era,[44] achieved a prominent position in iron metallurgy at this time.

During the seventh century A.D. a number of historical events occurred which had a profound effect on iron metallurgy. The forces of Islam established a secure hold on large areas of the countries between China and the Atlantic, thus carrying the techniques of Indian and Persian iron-making to the west; the widespread popularity of damascene steel marks this expansion. At the same time, Europe began to emerge from the post-Roman "Dark Ages," an emergence accompanied by a revitalization of iron industries. Moslem influence on European iron-making at this time is reflected in the Merovingian and Viking pattern-welding process—a laborious technique for the imitation of the appearance of damascene steel, requiring considerable skill in forging. At the same time, most of the techniques of producing steel of high quality became established; by A.D. 800 the direct process had reached its full development in Europe, and changed but little subsequently.[45]

42. Angus, Brown, and Cleere 1962.
43. Aitchison 1960, pp. 220–36.
44. Rickard 1932, pp. 863–65.
45. Aitchison 1960, pp. 236–59.

The History of the Indirect Process

A highly significant aspect of Iron Age technology, generally overlooked by archaeologists, is contained in the history of cast iron and the indirect process. The early development of this process can be said to have taken place almost exclusively within the borders of China, where the history of iron metallurgy followed a radically different course from that of the countries to the west.

Although it has been suggested that iron may have been known in China during the Shang Dynasty (1776–1122 B.C.), bronze continued to be the only important metal throughout the Shang, Western Chou (1122–722 B.C.), and early Eastern Chou or Ch'un Ch'iu (722–480 B.C.) periods. The use of bronze was monopolized by the aristocracy, and it found its primary application in the manufacture of military and ceremonial items; agricultural implements were generally made of wood and stone.[46] Toward the end of the Ch'un Ch'iu period, probably during the sixth century B.C., the evidence for the production of iron in China becomes conclusive. During the succeeding Warring States period (480–221 B.C.) iron occurs in abundance in the archaeological record. While firm evidence for the production of iron in China is, therefore, relatively late, wrought iron and cast iron appear simultaneously from the start. Iron objects from graves of the Warring States period in Manchuria, Korea, and North China have been shown to include both wrought iron and cast iron specimens;[47] a passage from *Tso Chuan* appears to describe the casting of iron cauldrons as early as 512 B.C.[48] It is therefore evident that the production of molten cast iron in China preceded the general use of this material in Europe by about twenty centuries; the under-

46. Kwang-chih Chang 1963, pp. 195–202.
47. *Ibid.*, pp. 196–97.
48. Needham 1958, p. 3.

lying causes of this early advance in iron metallurgy deserve careful scrutiny.

It is possible that the early Chinese iron smelters may have known how to produce iron in a bloomery furnace; however, the evidence for it is entirely lacking. The blast furnace seems to have been the mainstay of Chinese iron metallurgy from the very beginning. It is significant that there seem to be no Chinese words for the reduction of iron ore in a solid form; the words for smelting, *chu* and *yeh*, refer to the production of iron in its liquid form.[49] It can therefore be assumed that the early occurrences of wrought iron were probably the result of the fining of cast iron; the fact that the Chinese words for cast iron (*senq t'ieh*) and wrought iron (*shu t'ieh*) mean "raw iron" and "ripe iron," respectively,[50] supports this view of the indirect process as the basic tradition of Chinese metallurgy. The technological and economic factors which made possible this early production of cast iron have been explored by Sekino[51] and Needham.[52] The most important of these are probably to be found in an early sophistication in the control of heat: Chinese artisans of the seventh century B.C. possessed great skill in the casting of intricate bronze objects and were able to fire pottery at high temperatures. Considering the availability of iron ores in China, Sekino suggests that the use of magnetite would have necessitated high smelting temperatures and would therefore tend to produce cast iron. This, however, is a logical contradiction; more reasonably, Needham suggests that the high phosphorus content of Chinese ores, which reduces the melting point of cast iron, would have aided in the ease of its manufacture. The initial production of cast iron was reinforced by the invention of double-cylinder bellows with reciprocating motion (fourth century B.C.), the double-acting

49. *Ibid.*
50. *Ibid.,* p. 10.
51. Kwang-chih Chang p. 197.
52. Needham 1958, p. 47.

(push-pull) cylinder bellows (second century B.C.), and the application of water power to these bellows in the first century A.D.[53] A reinforcing result, though hardly a cause for invention as Sekino would have it, is the fact that the rendering of cast iron from ore allows for the mass production of the full range of alloys at a highly efficient rate. Thus, iron became the metal of the Chinese peasantry and found its early and large-scale use in the fashioning of agricultural implements. Iron axes, adzes, chisels, spades, sickles, and hoes were common implements of the Warring States period.[54] In the West, by contrast, iron was far beyond the means of commoners during the early centuries of its production.

While the consumption of cast and wrought iron for agricultural tools grew rapidly in China, bronze continued to be used for weapons for some time. This too came to an end when the crucial technique by which steel is refined from cast iron through decarburization was mastered.[55] Known as "the hundred refinings," this difficult process involved the repeated heating of a cast iron bar in a forge, and the judicious application of an oxidizing cold blast. "Great iron" (*chü t'ieh*), or steel, is mentioned as standard equipment of the army by about 250 B.C.; several steel spearheads and swords from this period have survived. The necessary techniques for hardening of the steel—quenching and tempering—were mastered during the Han Dynasty (221 B.C.–A.D. 220). The production of wrought iron from cast iron by prolonged roasting in an open hearth nevertheless required much less control than "the hundred refinings"; by the fourth century A.D., the simpler process found further use through the invention of co-fusion steel. By immersing heated wrought iron in molten cast iron, the carbon content of the two could be averaged and steel produced. Further technological advances involved the welding of hard and soft

53. *Ibid.*
54. Kwang-chih Chang 1963, p. 197.
55. Needham 1958, pp. 6ff.

steels for quality sword blades (third century A.D.), a process which was later brought to perfection in Japan. The importation of Indian *wootz* steel, started in the sixth century A.D., further added to the sophistication of Chinese swordmaking.

Between the fifth and seventeenth centuries, China enjoyed an abundance of iron of all types which was unknown in Europe. The twenty-foot high cast iron statue of a lion at Tschang-chou, erected in A.D. 954, is one of the world's largest iron castings; the thirteen-story cast iron pagoda of Yü Ch'üan Ssu at Tang-yang in Hopei, built in A.D. 1061, required some fifty-three tons of iron. The amount of metal in these ceremonial constructions, presumably amassed through some form of taxation, provides a clear indication of the capacity of the iron industry in medieval China.

In Europe, by contrast, the bloomery furnace and the direct process continued to be used almost exclusively until the late fourteenth century. This state of affairs was perhaps not so much due to the fact that cast iron could not be made, but rather that it was rejected as unsuitable when it was accidentally produced. Some evidence exists for the production of cast iron in Roman times, although this material seems not to have been used in the manufacture of usable objects. An inventory of Roman cast-iron finds, compiled by Coghlan,[56] contains at least three indisputable finds from England. A small block of coal-smelted, white cast iron (3.23 percent carbon, 0.485 percent sulfur) has been excavated at the iron smelting site of Wilderspool, which had probably been used by the twentieth Roman Legion. Charcoal-smelted examples exist in the form of a cast-iron bar from Tiddington (3.52 percent carbon, 0.035 percent sulfur), and a nondescript lump of metal from Hengistbury Head (3.49 percent carbon, 0.035 percent sulfur). Although Coghlan accepts these finds as at least some kind of evidence that Roman cast iron objects were made, the specimens in question were obviously not utilized. Cast iron statuettes of

56. Coghlan 1956, pp. 76–79.

alleged Roman age have been described (one from Sussex, two from Germany), but these items cannot be conclusively associated with Roman deposits. It appears, therefore, that the Romans were capable of producing cast iron, but that they were not able to make use of it. A few centuries later (*ca.* A.D. 800) there is a reference in a metallurgical manuscript which may be interpreted as a description of cast iron production in Europe,[57] but the evidence is not clear. The direct process continued as the mainstay of European iron production until the fourteenth century.

The thirteenth and fourteenth centuries in Europe saw the development of a number of metallurgical techniques which were to give rise to an iron industry based on the indirect process.[58] The Osmund furnace, developed in thirteenth century Sweden, was about ten feet high and received its blast from mechanically driven bellows. Furnaces of this kind increased the output of a single charge considerably and were adopted all over Europe. In 1311, German smelters developed the progenitor of the blast furnace: the *Stückofen* (also called the *Hochofen* or *Blauofen*) was about fifteen feet high, had water-powered bellows, and could produce up to a ton of bloomery iron per day. By 1340, the height of the German furnaces had grown to about thirty feet, and the blast pressures had been increased considerably; by 1380, the first true blast furnace for the production of cast iron was built. Cast iron in quantity was the result; for the first time in Europe this form of iron was put to use. The accompanying techniques of the indirect process followed quickly, and the foundations of the massive iron and steel industry of the West had been laid.

The adoption of the indirect process in the West is often held to be independent of the earlier invention of this process in China. Certainly, there is a clear evolutionary development of the blast furnace in thirteenth and fourteenth century Europe. Needham, however, points out a number of

57. Johannsen 1933.
58. Aitchison 1960, pp. 336–43.

facts which build a strong case for stimulus diffusion from China.[59] A number of "new" inventions of fourteenth century Europe had previously been developed in China: these include mechanical clocks, which first appear in Europe in 1310, gunpowder (1325), the segmental arch bridge (1340), and block printing (1375). It seems reasonable to assume that knowledge of the production of cast iron, but especially of the associated techniques for the production of wrought iron and steel, also diffused from China.

Toward Modern Iron Metallurgy

The introduction of the indirect process was soon to revolutionize the iron industry in Europe. By the end of the sixteenth century, blast furnaces were in operation in most European countries; in England, iron production had so depleted timber resources that an Act of Parliament in 1584 restricted the growth of the industry.[60] Sweden, with its natural abundance of forests, and with the stimulus of its involvement in the Thirty Years War, rose to the fore as the largest supplier of iron in seventeenth century Europe. In the United States, a strong rival industry began its development at such centers as Saugus, Massachusetts. By the end of the seventeenth century, world production of iron had risen to about 100,000 to 150,000 tons per year, while production in England had dropped to about 18,000 tons. The strain on the fuel supplies of other countries was also beginning to show. The situation was overcome by the successful use, in 1709, of coke for the smelting of iron. Used at first only at Coalbrookdale, the new process spread through England over the next forty years.[61] When the Industrial Revolution started changing the face of British industry by 1750, the iron industry was no longer hamstrung by a lack of fuel.

A number of crucial metallurgical processes (essentially

59. Needham 1958, p. 39.
60. Parr 1958, pp. 72–84.
61. Schubert 1957, p. 337.

industrial adaptations of old techniques) developed in eighteenth and nineteenth century England and proceeded to set the production of iron on its modern course.[62] Cort's reverberatory furnace, in which the puddling process was carried out, was patented in 1784. In this furnace, cast iron was melted by drawing hot gases over the metal from an adjoining chamber and oxidized by adding iron oxide. Neilson's hot air blast, patented in 1838, used preheated air to fan the furnace fire and reduced coke consumption from six tons to two tons per ton of cast iron produced. Finally, the Bessemer process (1856) and the Siemens-Martin open-hearth process (1861) of mass-producing steel at a highly efficient rate paved the way for the iron industry of today, which produces well over three hundred million tons per year.

Fuel: Charcoal *vs.* Coal

While the use of coked coal for smelting revolutionized the iron industry in England, and subsequently in the rest of the industrialized world, coal had been used for smelting in earlier times. Since the type of smelting fuel employed in the manufacture of an iron artifact becomes crucial when that artifact is to be dated by the carbon-14 method, the history of fuels in iron metallurgy is treated here in a separate section.

The quality of the fuel used in iron smelting determines the quality of the product; not only does fuel quality determine temperature, but extraneous materials in the fuel will affect the characteristics of the iron. Charcoal, although not capable of the high temperatures produced by coal, has a low sulfur content and is therefore better suited to primitive smelting processes. There are also some examples of primitive bloomeries in which wood is used directly as fuel; such a practice was observed in the Kordofan area in modern times.[63] Some woods, such as mimosa, acacia, or tamarisk

62. Parr 1958, pp. 96–128.

63. Forbes 1950*b*, p. 106.

can produce an 850°C fire without a blast and could certainly be used for smelting.[64] Charcoal, however, is a better all-purpose fuel and has been used from the earliest times for a variety of purposes. Storerooms in Egyptian tombs of the First Dynasty (*ca.* 3100 B.C.) have been found to contain supplies of prepared charcoal; in fact, the Egyptians had a special term and a measure for charcoal.[65] It can therefore be assumed that when the knowledge of iron-working was acquired, charcoal was immediately used to smelt it and continued to be so used through most of its history.

Not only is charcoal a very suitable fuel for smelting operations, but the charcoal from specific kinds of woods gives the best results. The best charcoal is prepared from woods with the sap still in it (i.e., green woods), which are slowly combusted in large heaps. That the ancients were well aware of this fact can be gleaned from a letter written by Hammurabi (*ca.* 1750 B.C.) to a servant, concerning the available charcoal supply.

> . . . Wood for . . . the metal-workers they shall seek for you in Dur-gûrgurri. . . . then they shall cut for you 7200 logs. . . . Every lot of 300 logs shall be loaded on a freighter and brought to Babylon. Among the firewood that will be cut, there shall be no wood that died in the forest. They shall cut green wood only.[66]

This knowledge of the characteristics of charcoal is ably explained by the Roman naturalist Theophrastus, in his *Enquiry into plants.*

> The best charcoal is made from the closest wood such as the holin oak, for these are more solid. . . . Worst of the woods mentioned is the oak, since it contains most mineral matter and the wood of older trees is inferior to that of younger, and for the same reason that

64. Coghlan 1956, p. 106.
65. Forbes 1950*b,* p. 106.
66. In Forbes 1950*b,* p. 107.

> of really old trees is especially bad. For it is dry, wherefore it sputters as it burns; whereas wood for charcoal should contain sap.[67]

While the techniques of producing charcoal may be of relatively slight historical significance, it is highly significant for the carbon-14 dating of iron. From the passages quoted, it is reasonable to deduce that throughout the history of iron metallurgy only freshly cut wood was generally used for the production of charcoal for smelting. In other words, there was no significant time lapse between the felling of the tree and the fueling of the furnace.

While charcoal continued to be used as a smelting fuel until modern times, the Romans and the Chinese used coal in the early centuries of the Christian era. Evidence for Roman use of coal has been found at several sites in England.[68] The coal-smelting of cast iron at Wilderspool has already been mentioned; at Tiddington, coal and iron slag occur in the same deposits and indicate at least some use of coal.[69] In China, coal has been used since the fourth century, perhaps longer, but apparently only in the crucible process. This practice "... permitted attainment of high temperatures in the large piles surrounding crucibles. At the same time these would exclude sulphur."[70] In this process "... the iron ore was reduced in batteries of elongated tube-like crucibles packed in a mass of coal."[71] Although documentary evidence is lacking to provide further details about the process, some form of carbon must have been placed *in* the crucibles with the ore, to impart the necessary carbon to the iron alloy.[72] Coal would contribute sulfur in such a situation; its addition to the ore would have defeated the purpose of the crucible process. It can, therefore, be assumed that charcoal or wood

67. *Ibid.,* pp. 108–9.
68. Fieldhouse, May, and Wellswood 1931, p. 15; May 1904, p. 76.
69. Coghlan 1956, pp. 107–8.
70. Needham 1958, p. 14.
71. *Ibid.*
72. Needham 1966: personal communication.

chips were used, much as in the Indian *wootz* process. If this were indeed the case, the carbon content of the iron produced in crucibles would not be derived from coal, and there should be no particular problem in dating it.

Following the abortive use of coal by the Romans, it was not seriously considered by Europeans until the British were forced to look for a substitute fuel for their dwindling charcoal supply. Experiments with coal in blast furnaces, however, produced highly unsatisfactory results. Not only does the coal contribute a high sulfur content to the metal at high temperatures, but it also cakes and prevents the charge from descending properly in the furnace. These problems were finally overcome by the coking of coal, much as charcoal is prepared from wood. The use of coke was adopted at most British foundries by 1750, but the process did not appear in the rest of Europe for some time. The first coke-fired furnace in France was built in 1785, in Silesia by 1789, and in Belgium in 1823.[73] The process was not adopted in the Ruhr until the 1850's, while the Swedish industry used charcoal until well into the twentieth century. In the United States, coke was first used in Pennsylvania in the 1830's. These dates only indicate the earliest use of the new fuel in various areas. Throughout the nineteenth century, coke-fired and charcoal-fired furnaces operated side-by-side. About 800 tons of charcoal-smelted iron per year were produced in Lancashire around 1839, while in 1873 the Newland furnace was being fueled with a mixture of four-fifths charcoal and one-fifth coke.[74] In addition, "coke iron" and "charcoal iron" were often deliberately melted together for specific purposes; the effect of such mixings on carbon-14 dates is demonstrated by some of the laboratory results reported in chapter 6.

73. Aitchison 1960, pp. 447–48.
74. Coghlan 1956, p. 107.

4

CARBON-14 DATING OF IRON

PRINCIPLES

MAN'S PRODUCTION of iron results in iron-carbon alloys of varying carbon concentration, depending on the kind of alloy involved (see chapters 2 and 3). It has been shown in the discussion of manufacturing procedures that the carbon content of the alloy derives from fuel used in the smelting furnace or carburizing forge and that this fuel, almost without exception, has been charcoal or coal throughout the history of the Iron Age. The possibility that the carbon-14 activity of carbon in iron alloys could be measured to provide a date of manufacture for a given object was first mentioned by Turekian.[1] The principles and assumptions involved in such a procedure will be discussed here in detail. The discussion will include an outline of the principles of carbon-14 dating; the characteristics of organic materials which are normally dated, and their effects on accuracy; the nature of iron samples, and the principles involved in their dating.

PRINCIPLES OF CARBON-14 DATING

Although carbon-14 measurements have been used for a wide variety of purposes, not all of which involve age deter-

1. Turekian 1961, p. 10.

minations, readers will be most familiar with their usefulness in the dating of organic materials such as charcoal, wood, peat, and the like, or inorganic material such as shell. The same theoretical principles regarding the production and decay of carbon-14, which are assumed in the dating of such materials, are of course involved in the dating of iron artifacts. Originally described by Libby, these principles have been thoroughly presented in a large number of primary and secondary sources;[2] only a brief review will be given here.

The carbon reservoir (atmosphere, oceans, and biosphere) contains carbon in the form of three isotopes: C^{12} and C^{13}, two stable isotopes, and C^{14}, a radioactive isotope. The relative abundance of the three isotopes differs somewhat between the biosphere and the inorganic parts of the reservoir (atmosphere and oceans), but the mean ratio between C^{12}, C^{13}, and C^{14} is stable at approximately 100:1:0.01. Of the three isotopes, C^{12} and C^{13} are not subject to physical changes, but the available supply of C^{14} decays radioactively at a constant rate of about 1 percent per eighty years. At the same time, C^{14} is continuously produced in the upper atmosphere by nuclear reactions associated with cosmic radiation. The highly energetic cosmic rays, arriving in the upper atmosphere from space, interact with atmospheric nuclei to produce free neutrons; colliding with nitrogen atoms, the neutrons form carbon-14.

$$N_7^{14} + n_0^1 = C_6^{14} + H_1^1$$

The radioactive carbon-14 atoms produced in this reaction have a half-life of 5,730±30 years, according to recent measurements;[3] their chemical properties, however, do not differ from those of the stable carbon isotopes. Thus, the carbon-14 atoms are almost immediately oxidized by at-

2. Libby 1952, 1955; Kulp 1952; Aitken 1961; Willis 1963.

3. Mann, Marlow, and Hughes 1961; Watt, Ramsden, and Wilson 1961; Ollson *et al.* 1962; Ollson and Karlén 1963; Hughes and Mann 1964. The "Libby half-life" value of 5568±30 years is still used, however, when carbon-14 age determinations are reported.

mospheric oxygen to form radioactive carbon monoxide molecules, which are in turn oxidized—either through direct reactions with atmospheric oxygen, or through bacterial action in the biosphere.[4] The result is a continuous supply of radioactive carbon dioxide entering the carbon reservoir.

The carbon reservoir consists of the atmosphere, the oceans, and the biosphere. Radioactive carbon dioxide becomes mixed with carbon dioxide in the atmosphere, thus providing a radioactive component in atmospheric carbon dioxide. Through interaction with surface water, radioactive carbon dioxide is dissolved in the oceans, where it is also absorbed as a component of bicarbonates and carbonates (e.g., in the shells of molluscs). In the biosphere, radioactive carbon dioxide is incorporated in plants through photosynthesis and is subsequently ingested, directly by herbivores and indirectly by carnivores. Thus, the atmosphere, the oceans, and all living matter receive a continuous supply of carbon-14 as a result of the cosmic ray bombardment of the upper atmosphere. The mean lifetime of a carbon-14 atom is about 8,000 years; the decay products of its disintegration are a nitrogen atom and a *beta* particle (electron).

$$C_6^{14} = N_7^{14} + \beta_{-1}$$

The production of carbon-14 in the upper atmosphere is in equilibrium with its radioactive decay from the carbon reservoir. For equilibrium to be maintained, it can be calculated that the total carbon-14 inventory of the carbon reservoir must be stabilized at about 81 metric tons.[5]

The equilibrium between carbon-14 production and decay is also maintained in each living organism. While the organism is alive and exchanging carbon with the atmosphere, its specific activity (carbon-14 activity per unit mass) will be the same as the carbon-14 activity of the atmosphere. When an organism dies, or when a tree ring has been formed, it no

4. Pandow, MacKay, and Wolfgang 1960; MacKay, Pandow, and Wolfgang 1963.

5. Libby 1955, p. 5.

longer acquires carbon dioxide from the atmosphere and its carbon-14 activity decreases through radioactivity decay. Since the rate at which this decay process takes place is known, a measurement of the carbon-14 activity of the organism can establish the date when it ceased to exchange carbon with the rest of the carbon reservoir. Such an age determination is made by comparing the activity of a given sample with the activity of modern organic material, according to the equation

$$t = -\lambda \log_e \frac{I}{I_0},$$

where t is the time since death, I the activity of the sample, I_0 the activity of modern material, and λ the decay constant of carbon-14. The best value of λ for a half-life of $5{,}730 \pm 30$ years is 8,267 years; for the "Libby half-life" of $5{,}568 \pm 30$ years, it is 8,033 years.

The principles of carbon-14 dating, as outlined above, are based on the assumption that the carbon-14 activity in the carbon reservoirs has been constant over time and that the half-life of carbon-14 is accurately known. Neither of these assumptions is strictly true, but it is ". . . a testimonial to Libby's original notion that the modifications, although important, detract little from the positive value of the method as a dating device."[6] Modifications are nevertheless required, and a number of investigators[7] have worked out the nature and extent of these problems; further corrections can be expected to appear in the future. Of the major problems which effect the underlying assumptions of carbon-14 dating, the following are crucial:

Half-Life

It has been mentioned in this chapter that recent measurements of the half-life of carbon-14 provide a value of $5{,}730 \pm 30$ years. By international agreement, however, the

6. Willis 1963, p. 35.
7. Suess 1965; Damon, Long, and Grey 1966.

original "Libby half-life" of 5,568±30 years[8] is still being used when carbon-14 dates are reported; this practice is followed in the present volume. To convert carbon-14 dates calculated from the "Libby half-life" to the new value, the age in years should be multiplied by 1.03. Further small corrections of the half-life value can probably be expected in the future.

Variations in Production of Carbon-14

The accumulation of carbon-14 age measurements for known-age samples (primarily tree rings) has provided a body of data which shows small but significant variations from the expected results.[9] The most important factor which accounts for these variations is the apparent variation over time of carbon-14 production. This phenomenon is linked to changes in the intensity of the cosmic-ray bombardment of the upper atmosphere and has not been fully explained, since the origin of cosmic rays is uncertain. Assuming, however, that cosmic rays are being produced at a constant rate, a number of factors have been shown to influence cosmic ray flux in the upper atmosphere. Both the earth's magnetic field and the magnetic field carried by charged particles emitted from the sun ("solar wind") deflect the cosmic radiation; changes in the intensity of these parameters result in variations of cosmic ray intensity in the upper atmosphere and hence in the amount of carbon-14 produced. Periods of increased sunspot activity result in a "solar wind" of higher intensity, with a corresponding decrease in cosmic-ray bombardment.[10] Similarly, an increase in the intensity of the earth's magnetic field will yield a decreased production of carbon-14.[11] Organisms living during a period of decreased carbon-14 production will exhibit carbon-14 ages somewhat

8. Libby 1955, p. 36.

9. de Vries 1958; Ralph and Stuckenrath 1960; Willis, Tauber, and Münnich 1960.

10. Stuiver 1961, 1965.

11. Elsasser, Ney, and Winckler 1956; Kigoshi and Hasegawa 1965.

older than the expected value; when there are decreases in the intensity of the earth's magnetic field, or in sunspot activity, the effect is reversed.

Carbon Reservoir Exchange Rates

Carbon-14 concentration in the atmosphere (and hence in the biosphere) is affected by changes in the exchange rates between the various parts of the carbon reservoir. Such variations may occur as a result of climatic changes. During a glacial period, for example, the oceans can keep less carbon dioxide in solution, and the carbon dioxide exchange between oceans and atmosphere is increased. At the same time, the mixing rate of the oceans is probably speeded up through weakening of the temperature inversion layer.[12] The result would be an upwelling of "old" carbon dioxide from the ocean depths, its subsequent mixing with atmospheric carbon dioxide, and a decrease in carbon-14 activity in the atmosphere.

A similar effect probably occurs when glaciers retreat and "old" carbon dioxide is released from the melting ice. The magnitude of these effects, however, is not accurately known.

Human Interference

Man has contributed his share in disturbing the carbon-14 activity of the atmosphere. Since the inception of the Industrial Revolution enough fossil fuel has been burned to dilute atmospheric carbon-14 activity by about 2 percent. Conversely, the nuclear explosions of the past twenty years have produced enough free neutrons to nearly double the atmospheric carbon-14 concentration. Future generations may find that the usefulness of the carbon-14 dating method ends in the mid-twentieth century.

The cumulative effect of the processes described above (excluding the "bomb effect") has resulted in atmospheric carbon-14 concentrations which deviate by 2 percent in

12. Suess 1965.

either direction from predicted values for the last 2,500 years. These fluctuations have been measured by comparing the carbon-14 dates of tree rings with their dendrochronologically determined ages. The results of the two dating methods, when plotted against each other, yield an undulating curve instead of a straight line.[13] A curve for the relation between carbon-14 ages and true ages for the past 1,000 years, provided by Stuiver and Suess, is reproduced in figure 2. From this curve it can be seen that a given carbon-14 age may correspond to more than one calendar age; a

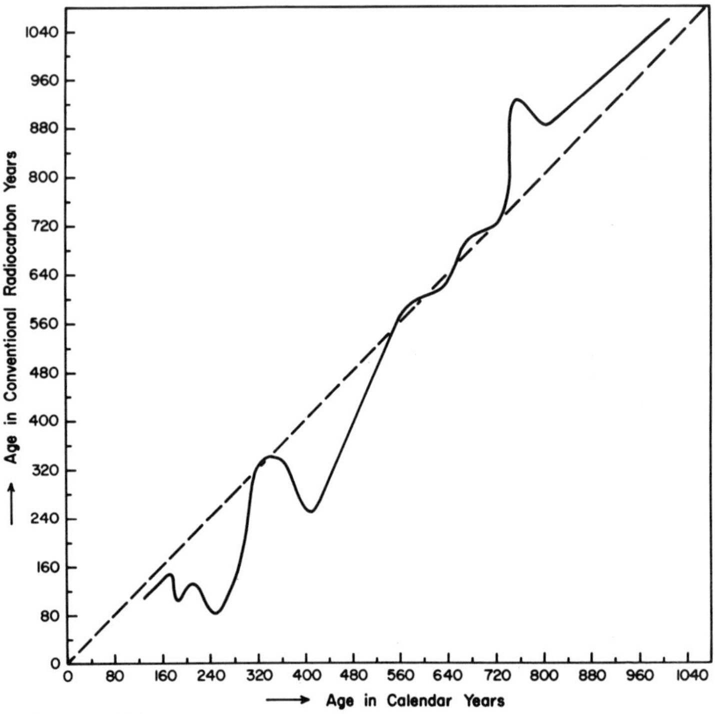

FIG. 2.—The relation between radiocarbon ages and true ages for the last millennium. (From Stuiver and Suess 1966, p. 538. Reproduced by permission of the authors.)

13. Stuiver 1961, 1965; Suess 1965; Stuiver and Suess 1966.

sample with a carbon-14 age of 120 years, for example, may have a calendar age of 140, 180, 200, or 220 years. A carbon-14 age of 560 years, however, corresponds to the actual age of the dated sample. Recent measurements have shown that carbon-14 ages of *circa* 4,000 years may actually be 500 years too young;[14] a number of laboratories are currently extending the calibration of carbon-14 ages back to several millennia B.C. through comparisons with tree rings and lake sediments.

Although the variation of carbon-14 ages from predicted values are significant, the variations are sufficiently small to allow for carbon-14 ages to be considered as approximate ages of samples, as long as the limitations are kept in mind. This is especially the case for the past 2,500 years, the period encompassing all the laboratory results reported in this book. Since the internationally established practice is still to equate carbon-14 ages and calendar ages, this practice will be followed, except where noted otherwise; the reader may correct carbon-14 ages by using the graph in figure 2.

The Carbon-14 Dating of Organic Materials

To gain a better understanding of the procedures involved in dating iron, and the possible errors which may arise as a result of the nature of the sample material, it is necessary to review the techniques by which organic materials are dated. Since the dating of iron involves the dating of charcoal which occurs in iron alloys, the discussion can be confined to the procedures employed in the conventional carbon-14 dating of charcoal.

Charcoal samples submitted for carbon-14 assays have generally been subject to a number of contaminating factors; these contaminants must be removed as thoroughly as possible before a sample can be dated. Rootlets, often microscopic in size, are a source of modern carbon in such samples and are mechanically or chemically eliminated. Carbonates

14. Stuiver and Suess 1966.

and humic acids, introduced to the sample by groundwater, are a source of old carbon and are decomposed by alternately leaching the sample with 2-percent solutions of hot HCl and NaOH and washing it with distilled water. The sample is then dried and is ready for analysis.

A variety of techniques exist to measure the radioactivity of carbon: the sample can be converted into carbon dioxide, methane, acetylene, or benzene for measurement in different kinds of counting equipment. Recent advances in the preparation of methane and in liquid scintillation counting techniques (used for detecting carbon-14 in benzene) have increased the popularity of the use of methane and benzene.[15] The majority of radiocarbon laboratories, however, use carbon dioxide as a counting gas: the sample is burned in oxygen to produce carbon dioxide and is then subjected to a series of chemical purification steps. The purified sample must contain less than one part in 1×10^5 of electro-negative impurities (mostly water vapor, oxygen, and chlorine) to be useful as a filling gas in a proportional gas counter. The counting procedure, which detects radioactive disintegrations in the carbon dioxide, usually lasts from twenty-four to forty-eight hours. The net activity of the sample (gross counts per minute minus background counts per minute) is then compared with the activity of a modern standard for an age measurement within statistical limits of accuracy. The background, which consists primarily of radioactive substances in the counting equipment and of cosmic radiation penetrating the counter shield, is characteristic of a given set of equipment; it is determined by measuring the activity of a sample of "dead" carbon like anthracite. The modern standard most commonly used is prepared in the form of oxalic acid by the U. S. National Bureau of Standards to simulate the atmospheric carbon-14 activity of the year 1950, as it should have been without the deviations introduced by the fossil fuel and nuclear bomb effects. An efficient set of counting equipment can measure carbon-14 ages

15. Dyck 1967.

of up to 55,000 years; by means of the laborious process of isotopic enrichment, large samples can be concentrated to allow for age measurements of up to 70,000 years.[16]

The carbon-14 age of a charcoal sample is of course only the age of a given piece of burnt wood. In archaeological chronologies, where the tolerance of age deviations is necessarily restricted, additional problems exist. The use of old logs for firewood, or the burning of thick logs of which the inner rings are of considerable age, erroneously increases the ages of associated archaeological finds. In addition, the demonstration of direct associations between charcoal samples and past human activities becomes critical.

The Carbon-14 Dating of Iron

The carbon content of an iron specimen—be it wrought iron, cast iron, or steel—can be assumed to be derived from the charcoal or coal used to smelt it; in the case of steel, part of the carbon content may be derived from the fuel used in the carburizing process. If the fuel used is coal, the carbon in the metal cannot, of course, provide a direct age measurement. The use of coal or coke in iron production, however, preceded the eighteenth century only in rare instances, and has been documented in earlier chapters; a maximum age can usually be assigned from the cultural content when an iron sample shows no carbon-14 activity. When the fuel is charcoal, however, the carbon in the iron alloy can be employed in the same manner as charcoal, as it is dated in the more conventional process. The procedures involved in the dating of iron differ from those used to date charcoal in two major aspects: The contaminants which effect iron samples are different from those which effect charcoal; the sample is therefore prepared for analysis in a different manner. The carbon in iron can also not so readily be converted to carbon dioxide as charcoal; the process of extraction is therefore differently accomplished.

16. Haring, de Vries, and de Vries 1958.

Charcoal which is buried in the ground is subject to contamination by rootlets, carbonates, and humic acid. The carbon in iron, on the other hand, is essentially sealed in a sterile environment at the time of manufacture, and is subject to far fewer sources of contamination. A number of contaminating agents, however, must be considered as possible contributors of modern carbon in the analysis of iron specimens: Iron corrodes easily under wet conditions when enough oxygen is present; most ancient iron specimens have at least a surface layer of iron oxide (primarily Fe_2O_3), while some have been totally converted to oxide. While the carbon in the specimen remains unaltered by the corrosion process, the porous nature of the corrosion products may allow for the inclusion of organic material in small quantities, especially at the surface. A further problem may result from the fact that a corroded surface adsorbs more atmospheric carbon dioxide than a clean iron surface. The adsorption of gas molecules to a solid surface is described by the equation

$$\sigma = \frac{Np\tau_0 e^{Q/RT}}{\sqrt{2\pi MRT}}$$

where σ is the number of gas molecules per unit area of the adsorbent surface; N is Avogadro's number; p is the pressure of the gas which is being adsorbed and M its molecular weight; R is Boltzmann's gas constant; Q is the heat of adsorption; T is the absolute temperature; τ_0 is the time of oscillation of a gas molecule in the adsorbed form.[17] With all other variables remaining constant, τ_0 (and hence σ) will be larger for an adsorbing surface of iron oxide than for a clean iron surface: the magnitude of τ_0 is directly proportional to the time of vibration of the molecules of the adsorbent which, in turn, is directly proportional to the square root of its molecular weight.[18] Hence the adsorbency of Fe_2O_3 is larger than that of pure iron in the proportion $\sqrt{160} : \sqrt{56}$, or approxi-

17. de Boer 1953, p. 45.
18. *Ibid.*

mately 2:1. In addition, a corroded specimen usually has a larger surface area than a clean one, allowing for further increases in the amount of atmospheric carbon dioxide adsorbed. The inclusion of organic materials in corroded iron can be eliminated, and the problem of adsorbed carbon dioxide reduced, by removing all corrosion from iron specimens. For surface-corroded samples, this is easily accomplished by dry peening ("sandblasting," normally using an abrasive of plastic particles), with the possible addition of a weak acetic acid leach. Adsorbed carbon dioxide can be further eliminated by heating the sample under vacuum and by extracting the datable carbon without exposing it to the atmosphere again. (In the equation above, $\sigma = 0$ when $p = 0$ or when $T = \infty$.) Specimens which are corroded throughout cannot be entirely freed of organic contaminants, but a thorough surface cleaning should reduce the problem to acceptable limits. In a series of experiments with different cleaning methods (chapter 6), it was found that contamination of the corroded specimens used, either by organic materials or adsorbed carbon dioxide, was negligible. The theoretical considerations which have been discussed nevertheless argue for the thorough cleaning of specimens. It can be confidently stated, however, that contamination problems play a far less important role in the carbon-14 dating of iron than in the dating of materials like charcoal.

The extraction of carbon from an iron alloy, which is described in detail in chapter 5, can be accomplished in a number of ways. In the present study, the established procedure has been to mill the sample into chips and then to combust it in oxygen at about 1,100°C. Nearly all the carbon in the specimen is oxidized at this temperature; the resulting carbon dioxide is then cleaned in much the same manner as is the case with charcoal samples, and its carbon-14 activity is determined for an age measurement.

A carbon-14 date, if obtained from an iron specimen, can differ from dates obtained directly from charcoal in a number of significant ways. It can be shown, both theoretically and experimentally, that iron specimens are not subject to as many possible sources of contamination; the laboratory results, therefore, can be assumed to be more consistently accurate. It has also been shown, from historical and technological facts, that only freshly cut wood has generally been used to produce charcoal for iron smelting. While the problem of older tree rings in the carbon sample may therefore still be encountered, it can safely be assumed that there has been no significant time lag between the felling of a tree and its use in the production of the iron specimen concerned. The problem of demonstrating an association between an archaeological assemblage and a carbon-14 date is also significantly reduced when a cultural item such as iron is dated. Taken together, these considerations clearly indicate the desirability of dating iron specimens in preference to materials like charcoal when the choice presents itself.

While the present research project has focused on the dating of iron specimens, attention has also been paid to the dating of the waste product of iron smelting—that is, slag. Described fully in chapter 7, this procedure differs but little from the treatment of charcoal samples. Cinder, which is formed in the bottom of smelting furnaces and which is practically indestructible, often contains visible nodules of charcoal; the same is also true for slag formed in the smelting of copper. The slag can be dated by prying the charcoal from the slag and treating it in the conventional manner. While the problem of contamination is the same as for charcoal samples, the dating of slag has many of the advantages of dating iron: the charcoal comes from wood cut at the time of smelting, while cultural associations are readily demonstrable. The method can also be a useful survey tool where smelting sites are involved, since little excavation is required to collect datable samples.

5

CARBON-14 DATING OF IRON

LABORATORY METHOD

THE LABORATORY PROCEDURES used in carbon-14 dating have been actively utilized and improved during the past fifteen years in a large number of carbon-14 laboratories, while new improvements (especially in the sophistication of counters) are constantly being made. These procedures are, with minor variations, standard for all laboratories which use a given extraction and counting system. The Yale Radiocarbon Laboratory uses carbon dioxide as a counting gas for proportional counting; this method, which was used throughout the present project for the dating of iron, is utilized by a majority of carbon-14 dating laboratories and will be described only briefly. The major deviation from standard procedures —the extraction of carbon from iron—will be discussed here in detail.

To date iron, carbon is extracted from the alloy as carbon dioxide, which is then purified and used as a counting gas. This extraction is accomplished by direct combustion at high temperatures after the specimen has been cleaned and milled. Given a specimen of sufficient carbon content, a sample of carbon dioxide can be recovered which will allow an age determination to be made after the gas has been purified. The procedures followed to obtain a date for an iron specimen are outlined here in discussions of the treatment of the sample before analysis; extraction of carbon from the sample and the combustion equipment used for this purpose; purification of the counting gas; and counting and age determination of the extracted carbon dioxide. A final section deals with the problems of obtaining sufficient carbon dioxide from an iron specimen, and how this is affected by the kind of sample involved.

Treatment of the Iron Sample

Most iron samples submitted for carbon-14 dating are extensively corroded. Wrought iron specimens generally suffer the most—the object is often converted to a flaky mass of oxide—while steel and cast iron are corroded at least at the surface. Carbon in the alloy remains virtually unchanged, but the corrosion products could be a source of contamination and should therefore be removed wherever possible. A variety of techniques can be employed to accomplish this end, depending on the degree of preservation of the sample.

A specimen which is still structurally sound, and corroded at the surface only, can best be cleaned by dry peening, using an abrasive of plastic particles of about 90 mesh. Corrosion products are readily removed, leaving a shiny, new surface. A weak acetic acid leach and washing with distilled water serves to remove any surface impurities which may have been introduced during dry peening. As a further precaution, the dry peening abrasive should be tested for radioactive contamination, if there is the slightest possibilities that such contamination could have been introduced. (In many university workshops, repairs to components of nuclear accelerators, and the like, may render the equipment mildly radioactive.) This problem can be avoided by testing a sample of the abrasive in a thin window counter, to ensure that only background radiation is present.

Specimens which are extensively corroded cannot be dry peened, while any extensive acid leaching will dissolve the sample. Such samples are best cleaned by using a weak leach of very brief duration, followed by extensive washing with distilled water. The sample can be wrapped in fine copper screening, boiled in water, and then placed under a steady stream of water. When a sample is so extensively corroded that this method of cleaning must be adopted, it is usually sufficiently friable to allow it to be crushed (with mortar and pestle, for example) or broken up with a pair of pliers and used directly for combustion.

Specimens which cannot be crushed are milled to obtain small chips for the combustion process. Once again, workshops which handle radioactive materials are best avoided. The sample can be rendered into chips of about 10 to 30 mesh, although smaller or larger chips need not be excluded. Larger pieces, however, do not burn as readily and may result in incomplete extraction; smaller particles tend to foul the purification equipment with iron oxide dust. The milling is carried out by using a high speed, carbon-steel end mill when relatively soft samples of regular shape are involved. For harder materials, a carbide end mill may be necessary, while a carbide fly tool is required for samples of small size or irregular shape. Lubricants cannot, of course, be used to facilitate the milling. The milled chips are recovered by carefully cleaning the machine before use and covering its surface with a sheet of clean plastic, aluminum foil, or smooth paper. Most of the chips will remain on this cover and can be poured into a bottle; stubborn particles are gathered with a magnet.

Carbon Extraction: Equipment

The procedure which was established for the extraction of carbon from iron alloys involves the combustion of the sample in a stream of oxygen at temperatures ranging from 900°C to 1,100°C. Since this combustion produces a certain amount of iron oxide dust, which is not readily filtered out and therefore contaminates the inside of the equipment, it was found convenient to build a combustion unit and a partial gas purification unit which could operate separately from the standard gas-purification system in the laboratory, and which could be easily taken apart and cleaned.

The combustion equipment used (figs. 3 and 4) was designed to accommodate up to 150 grams of milled iron in a single charge and consists of two combustion units operating in series. Vacuum-tight components of this size for the combustion of iron are not readily available, and the system is therefore described in some detail: further information is provided in appendix 1. Alternate designs are of course possible and several modifications to the present system have been suggested by various researchers.

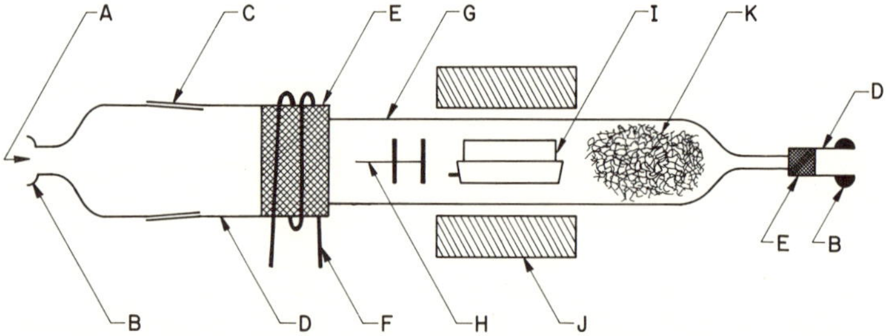

FIG. 3.—Combustion apparatus (only one unit shown): *A*—oxygen inlet; *B*—ground ball/socket joints; *C*—ground joint; *D*—graded Pyrex-Kovar seal; *E*—Kovar component, wrapped in copper screen, attached with epoxy resin; *F*—water-cooled jacket of copper tubing; *G*—zircon combustion tube; *H*—porcelain baffles wrapped in aluminum foil; *I*—milled sample in covered quartz combustion boat with RR Alundum bedding; *J*—electric furnace; *K*—quartz wool dust filter.

The combustion train is fed from a carbon dioxide–free oxygen supply which is controlled by means of a gas flowmeter (0–1.4 liters/minute) and connected to the system through a high vacuum stopcock and a series of ground joints. Two zircon combustion tubes (2″ O.D., 1–¾″ I.D., tapered at one end), capable of withstanding very large tem-

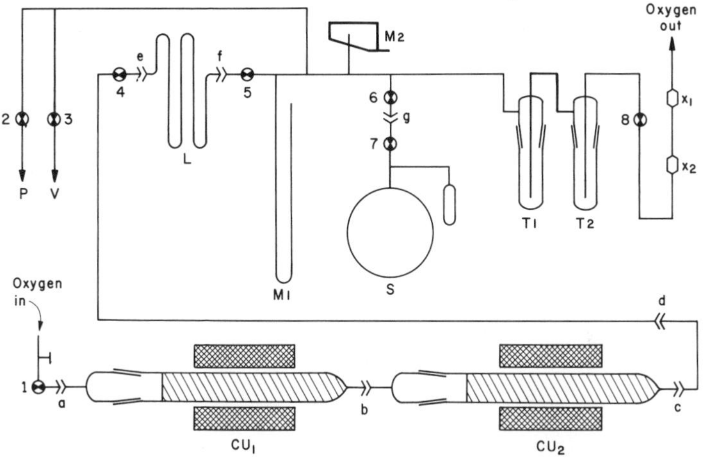

FIG. 4.—Combustion train assembly and partial gas purification unit; 1–8—Stopcocks; a–g—ground ball/socket joints; $CU_{1\text{-}2}$—combustion units; *P*—rotary oil forepump; *V*—high vacuum mercury diffusion pump; *L*—coil, at —80°C, for condensation of water vapor; M_1—0–100 cm mercury manometer; M_2—McLeod high vacuum gauge; *S*—detachable bottle with freezing trap, for storing CO_2; $T_{1\text{-}2}$—freezing traps (liquid nitrogen) for separation of CO_2 and O_2; $X_{1\text{-}2}$—ball valves filled with oil.

perature variations, are connected in series for sample combustion. Each combustion tube is heated by an electric furnace, of which the temperature can be controlled to its maximum of about 1,150°C. Vacuum-tight Pyrex-Kovar seals are attached to the ends of the tubes with epoxy resin, allowing the tubes to be connected to each other and to the rest of the equipment by means of Pyrex ground joints. To protect the epoxy seals from furnace heat, fine copper screening is wrapped around each Kovar component to dissipate heat, and water is circulated around the two wide seals in a coil of ¼ inch tubing. Porcelain baffles, wrapped in aluminum foil, are placed inside the combustion tubes near the wide ends to reflect radiant heat; quartz wool plugs, placed

FIG. 5.—Equipment for the extraction of carbon from iron alloys and for partial purification of carbon dioxide.

inside the tubes near the narrow ends, filter out iron oxide dust and shield the smaller seals from radiant heat. The milled sample is placed inside the combustion tubes in a pair of quartz combustion boats with covers, lined with 90 mesh RR Alundum (Al_2O_3 with less than .0015% carbon) as bedding material. These combustion boats, measuring 10 inches in length and 1–10/16 inches in diameter, were custom made to accommodate up to 75 grams of milled iron each.[1] Completing the combustion apparatus, Corning sil-

1. Combustion boats of refractory materials like zircon could not be obtained commercially in this size but would be an improvement since quartz cracks easily after large temperature changes.

icon grease is used to seal the ground joints around the combustion tubes; all other ground joints in the apparatus not subject to furnace heat are sealed with Apiezon "N" high-vacuum grease.

The combustion equipment is connected to a high-vacuum system (fig. 4) in which carbon dioxide can be partially puri-

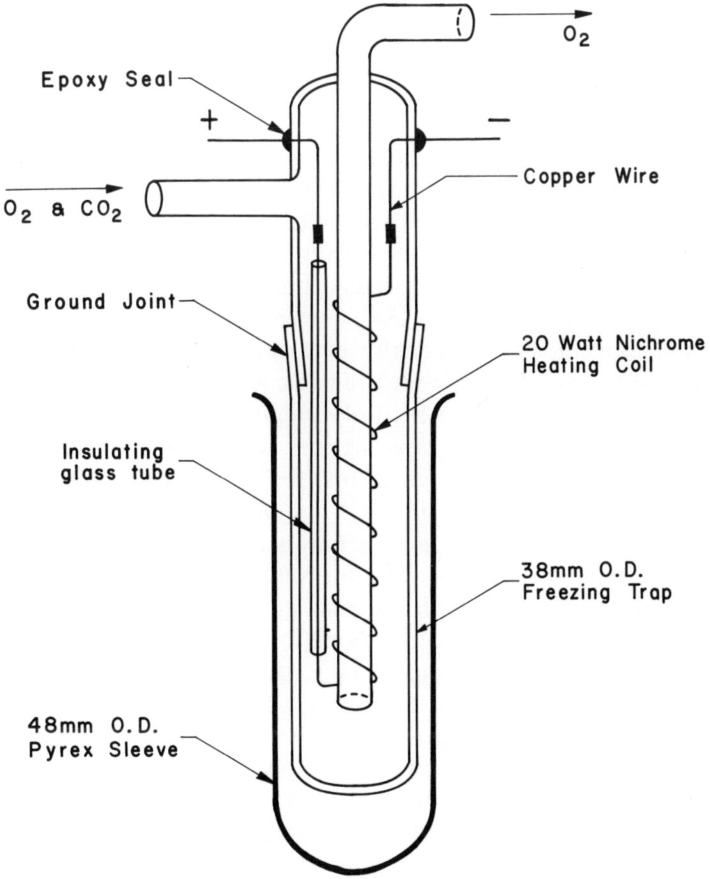

FIG. 6.—Detail of freezing trap for separation of CO_2 and O_2, using liquid nitrogen coolant.

fied and separated from excess oxygen. This system includes a detachable coil (L) of 10 mm Pyrex tubing, cooled by a mixture of acetone and dry ice to eliminate water vapor, and a pair of detachable freezing traps (T_{1-2}) in which carbon dioxide and oxygen are separated through fractional distillation. Since carbon dioxide freezes below $-78.5°C$, and oxygen liquifies below $-183°C$, carbon dioxide can be trapped and oxygen allowed to flow through if the temperature in the freezing traps is kept slightly above the oxygen boiling point. The boiling point of liquid nitrogen, the cooling liquid, is $-195.8°C$; the temperature inside the freezing traps must therefore be raised more than 13°C to allow oxygen to flow through. This is attained (fig. 6) by partially insulating the freezing trap walls from the liquid nitrogen by means of loose Pyrex sleeves, and by heating the traps from the inside with heating coils of 20 gauge nichrome wire. The coils, which pass through small epoxy-sealed holes in the freezing trap walls, are connected to a D.C. power supply; optimal power consumption, experimentally determined for the traps in use, was about 40 watts for the two coils (fig. 7).

With this arrangement, only small quantities of liquid oxygen gather in the traps; oxygen which flows through is allowed to escape through a "bubble" line (X_1–X_2 in fig. 4) consisting of Tygon tubing and two plastic ball valves filled with oil. The high vacuum system is completed by a 0–100 cm mercury manometer (M_1) for reading the pressure in the system; a McLeod high vacuum gauge (M_2) for reading pressures below 1 mm; a rotary oil forepump (P) and a mercury vapor diffusion pump (V), capable of producing a vacuum in the order of 10^{-4} mm; and a detachable three-liter round-bottomed flask (S) with a freezing trap, in which carbon dioxide can be trapped and stored. When the apparatus is being used for combustion, oxygen enters the system through the flow-meter, flows through the combustion tubes, the water vapor trap, and the two freezing traps, and bubbles out through the ball valves. When a high vacuum is required during the final stages of carbon extraction, the combustion

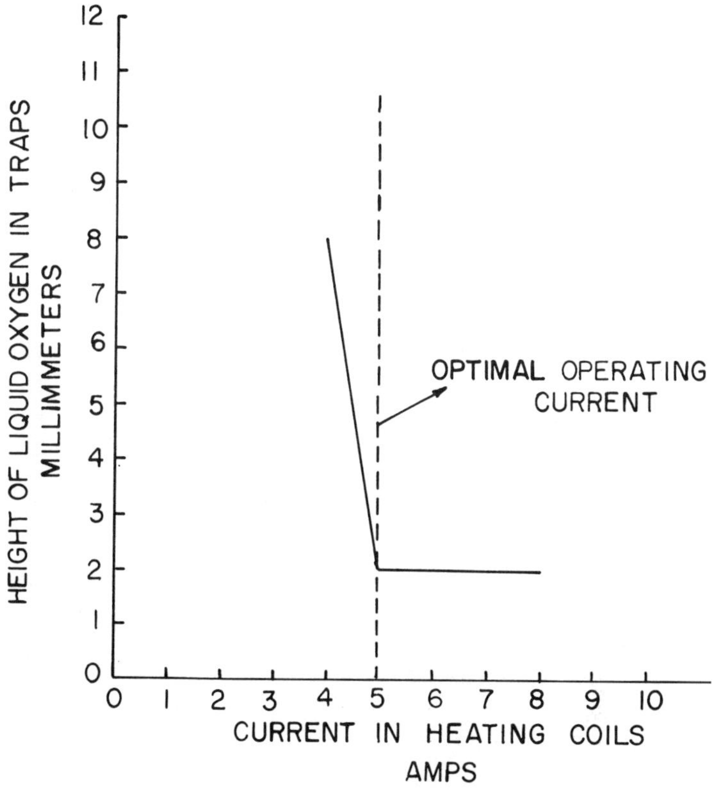

FIG. 7.—Liquid oxygen collected in freezing traps as a function of current in heating coils. Oxygen flow 0.5 liters/minute for 30 minutes.

apparatus, water vapor trap, and oil bubble line are shut off by closing stopcocks at 5 and 8 and only employing the high vacuum system.

Carbon Extraction and Partial Gas Purification

After the iron sample is milled, it is weighed and used to fill one or two Alundum-lined combustion boats, depending on the amount of sample available. The combustion boats are covered and inserted in the combustion tubes, heat shields are put in place, and the ground joints greased and closed. The entire system is then evacuated and the pumps left running for about 30 minutes to an hour to remove adsorbed carbon dioxide from the sample. (In evacuating the combustion tubes, pressures of about 1 micron were found to be attainable.) After pump-down, the combustion apparatus and water vapor trap are shut off from the high vacuum system and filled with oxygen. The high vacuum system and the storage vessel are then further evacuated until no pressure is indicated on the McLeod gauge (i.e., a vacuum of about 10^{-4} mm Hg is obtained) and checked for those ubiquitous leaks which tend to develop from time to time. The pumps and the storage vessel are now shut off and the combustion equipment, as well as the high vacuum system, are flushed with oxygen for about 10 minutes while preparations are made for the combustion process. The oil bubble line, which provides sufficient resistance to the oxygen flow to create an overpressure of 2 cm Hg in the apparatus, prevents atmospheric gases from entering the system from this point on.

When the system has been flushed with oxygen, a Dewar flask containing a mixture of acetone and dry ice is placed around the water vapor trap; Dewar flasks with liquid nitrogen are placed around the sleeved freezing traps; the heating coils in the traps are switched on; and the oxygen flow is adjusted to keep the ball valves bubbling intermittently (about 0.2 liters/min.). When the oxygen flow has been stabilized, the first furnace in the combustion train is switched on at full power. The reaction which follows depends largely on the size of chips to which the sample has been reduced (small particles burn at lower temperatures) and the alloy composition of the material (cast iron burns at lower temperatures). Generally combustion starts after the furnace has heated up to about 900°C to 1,000°C; oxygen stops flowing through the bubble line and the manometer indicates a loss of pressure. The flowmeter is adjusted to maintain intermittent bubbling through the ball valves. For particularly rapid combustion, oxygen consumption may rise to more than 1.4 liters/minute for several minutes, while most samples consume 0.5 to 1.0 liters/minute for about 5 minutes, and then slowly return to the normal flow of 0.2 liters/minute. The combustion process, which lasts about 10 minutes, raises the temperature inside the combustion tube sufficiently to fuse most of the iron and the bedding material together (probably in the 1,400°C to 1,500°C range); once combustion is proceeding rapidly, the furnace power supply may be reduced and the furnace temperature allowed to stabilize. When it is obvious that the combustion is completed, the first furnace is switched off, the system flushed with oxygen for a few minutes, and the entire procedure repeated with the second furnace.

The products of the combustion process include the desired carbon dioxide plus water vapor, finely divided iron oxides (mostly Fe_2O_3), and the oxides of many of the elements present in the iron alloy, carried in the stream of excess oxygen. Most of the iron oxides and water vapor are removed by the quartz wool plugs and water vapor trap, respectively, (although small amounts of both end up in the freezing traps), while most of the oxygen escapes through the bubble line. The final content of the freezing traps is a rather impure sample of carbon dioxide (dry ice) with admixtures of liquid oxygen, frozen water vapor, and traces of a variety of impurities. The high vacuum system is now closed off, the heating coils in the traps switched off, the sleeves removed, and the Dewar flasks of liquid nitrogen replaced around the traps. By evacuating the system to a pressure of less than one micron, most of the remaining oxygen is removed from the sample. The Dewar flasks are now removed and the sample allowed to sublime. If the manometer reading indicates that a sufficient volume of carbon dioxide has been collected to allow an age determination, the sample is trapped in the storage vessel, which can then be removed.

Gas Purification and Counting

After treatment in the system described, the sample normally contains impurities of up to 25 percent of sample volume, consisting of water vapor, some oxygen, and small quantities of sulfur oxides, nitrogen oxides, chlorine gas, and other substances. Final purification of the sample is accomplished by means of the standard gas purification system used in radiocarbon laboratories.[2] The apparatus used for this purpose in the Yale Radiocarbon Laboratory is essentially the same as the system described by de Vries and Barendsen[3] and is presented here only in outline form. The various chemical steps involved are designed to remove electro-negative impurities from the sample gas, which may render it unsuitable in the counting procedure.

The storage bottle of impure carbon dioxide, collected from the combustion of the iron sample, is connected to the standard gas purification system, and the gas is once more carried through the apparatus in a stream of oxygen. The procedure described in the preceding section is therefore partially duplicated, with the important difference that no iron oxides are present to foul the equipment. The major steps involved in purification are as follows:

1. Hot copper oxide (CuO) serves to oxidize any carbon monoxide which may be present as a result of incomplete combustion, thus adding to the volume of the carbon dioxide sample.[4]
2. The sample gas is bubbled through potassium perman-

2. Sample purification and counting were carried out by the staff of the Yale Radiocarbon Laboratory, under the direction of Minze Stuiver.

3. de Vries and Barendsen 1953; de Vries 1955; Barendsen 1955.

4. The oxidization of carbon monoxide by CuO properly belongs in the combustion train assembly, although it was not included in the experiments described.

FIG. 8.—Combustion and gas purification system in the Yale Radiocarbon Laboratory, normally used for the preparation of carbon dioxide from organic materials for carbon-14 dating. (Courtesy Minze Stuiver.)

ganate ($KMnO_4$) solution to oxidize sulfur dioxide and other impurities, facilitating some of the chemical steps which follow.

3. Repeated flushings through heated silver wire precipitates sulfur oxides as silver sulfates and also removes chlorine gas.
4. A series of cooling traps (of the same type as described in the preceding section) removes water vapor and allows oxygen to escape from the system.
5. The sample gas is now highly purified, but may still retain traces of electro-negative impurities, some of them unknowns. In the final purification step, the carbon dioxide is combined with calcium oxide (CaO) at about 800°C to produce calcium carbonate ($CaCO_3$). The resulting mixture of $CaCO_3$ and excess CaO is then cooled to about 450°C and pumped to high vacuum for about an hour. The carbon dioxide is driven off again at about 900°C, leaving the impurities behind. The CaO can be cleaned for further use by pumping it to high vacuum at a higher temperature.
6. The carbon dioxide is trapped and stored for about a month to allow radon—a chemically inert but radioactive gas which is naturally present in most materials—to disintegrate.

When the sample is ready for counting, it is used to fill a proportional gas counter to a pressure of about three atmospheres; counting time ranges from twenty-four to forty-eight hours, depending on the size and age of the sample. For this purpose counters with internal volumes of 0.5 liters (background 0.6 counts/minute) and 1.5 liters (background 1.2 counts/minute) were employed.[5] A mixture of 90 percent argon and 10 percent methane was used as a counting gas in the guard ring of Geiger counters, and the counting unit mounted inside a shield of armor plate and 200-year-old lead.

5. These highly efficient, low-background counters were designed and constructed by Minze Stuiver and will be described by him in a forthcoming publication.

FIG. 9.—Carbon-14 counting equipment in the Yale Radiocarbon Laboratory; the counter shield is open, showing three counters. (Courtesy Minze Stuiver.)

PROBLEMS OF CARBON DIOXIDE YIELD

While counters which require as little as 0.2 liters of carbon dioxide (equivalent to 100 mg of carbon) are presently being used in a few laboratories, such small samples require inordinately long periods of counting before a reasonably accurate measurement can be attained; most laboratories prefer to use sample of 5 to 10 grams of carbon. The smaller counter used for the present project normally required 1.9 liters of carbon dioxide, the equivalent of 1 gram of carbon (24 to 48 hours counting time), while longer measurements at lower pressures allowed samples as small as 0.5 grams. Generally, samples larger than 1 gram are preferable; this fact becomes extremely important in the selection of iron samples for dating.

The effects of sample size on accuracy in carbon-14 dating can be illustrated by calculating the standard measurement error (the familiar ± figures of carbon-14 dates) for a series of samples of different sizes. Assume that it were possible to construct an ideal counter, which has a 100 percent counting efficiency, and which has no background. Since recent carbon has a specific activity of about 13.6 disintegrations per minute per gram,[6] such a counter would register about 19,600 counts in twenty-four hours if it were filled with the carbon dioxide produced in the combustion of 1 gram of "modern standard" carbon. The standard counting error Δn is calculated from the equation

$$\Delta n = \frac{1}{\sqrt{n}}$$

where n is the total number of disintegrations registered by the counter. For a one-gram sample, therefore, the counting error is about 0.7 percent, which corresponds to a carbon-14 age error Δt of about sixty years. The results of similar calculations for samples of different sizes are given in table 1 and are graphed in figure 10. It can be seen that the standard deviation increases rapidly as the size of the sample decreases, with the deviation striving toward infinity as the sample size approaches zero. It should be emphasized that these calculations are for a hypothetical counter only; a real counter is not 100 percent efficient, has a background, and cannot measure backgrounds or comparative standards with perfect accuracy. These factors add about 50 percent to the standard error for samples over 1 gram, while the additional error becomes progressively larger for smaller samples, as low counting rates are "drowned out" by the counter background. In practice, the standard error for a 0.01-gram sample is already infinite.

6. Karlén *et al.* 1964.

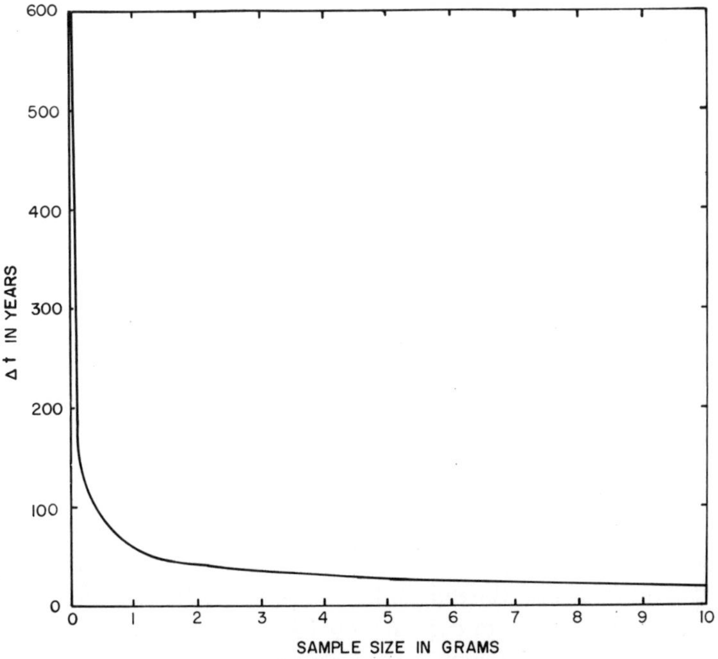

FIG. 10.—Carbon-14 age error as a function of sample size, calculated for 24 hour measurement in an ideal counter.

TABLE 1

STANDARD DEVIATIONS OF C^{14} MEASUREMENTS IN AN IDEAL COUNTER AS A FUNCTION OF SAMPLE SIZE (Rounded figures)

SAMPLE SIZE (GRAMS)	n; CALCULATED NUMBER OF COUNTS IN 24 HOURS	$\sqrt{n}$	$\frac{1}{\sqrt{n}}$ (Δn)	Δt IN YEARS
10	195,800	442	.002	20
5	97,900	312	.003	30
1	19,600	139	.007	60
0.5	9,800	98	.010	80
0.2	3,900	62	.016	130
0.1	1,960	44	.023	180
0.01	190	13	.071	600

A carbon sample of 1 gram can, theoretically, be extracted from a cast iron specimen weighing 50 grams, if the carbon content of the metal is 2 percent. The average loss during milling for samples of this size is about 30 percent, while the carbon extraction process is probably only about 90 percent efficient. The specimen must therefore actually weigh about 80 grams for successful analysis. More is lost for smaller samples; a 50-gram sample with a carbon content of 2 percent would probably provide 0.5 grams of carbon. Cast iron, however, often contains more than 2 percent carbon and smaller samples can therefore be dated; for steel and wrought iron, with much lower carbon contents, the specimen must be correspondingly larger. The size of actual samples which were successfully dated during the research project varied from about 35 grams to 110 grams for cast iron, while a bloomery iron sample of 600 grams (carbon content 0.2%) and a steel sample of 220 grams (carbon content 0.66%) also yielded sufficient carbon dioxide (see appendix 2).

To determine the suitability of an iron specimen for dating, its total carbon content must be known. A carbon content analysis of about 1 or 2 grams of the sample can be made by means of a simple combustion system;[7] if a laboratory has to make such determinations regularly, a carbon analyzer can be employed (see appendix 1). It is also possible, of course, to estimate the amount of carbon in a specimen if its alloy composition is known; alloy type can often be guessed with fair accuracy on the basis of the provenience, function, appearance, and presumed age of the specimen. With the carbon content known, sample suitability can be judged by calculating the amount of carbon in about 60 percent of the specimen. If the sample falls short of the required limit by only a small margin, it can sometimes be salvaged by crushing it to avoid milling loss, although this procedure complicates the combustion process.

7. Welcher 1963, pp. 654–59.

6

CARBON-14 DATING OF IRON

EXPERIMENTAL RESULTS

THE PROCEDURES for the carbon-14 dating of iron were tested experimentally in fifteen instances through carbon-14 measurements. The results of these determinations confirm the theoretical expectations that iron can be dated accurately by the carbon-14 method, and serve to illustrate the major problems involved in such a procedure. The results fall roughly into three groups, depending on the types of samples involved; these are summarized below and outlined in table 2, while a detailed account of the analysis of each sample follows in the rest of the chapter.

The first group of samples was used to check the operational qualities of the extraction equipment, to determine the extent to which contaminants may effect the accuracy of the results, and to establish the best procedure for cleaning iron samples. A sample of anthracite (Y–1500), combusted instead of iron, provided a carbon-14 activity measurement of 0.0 ± 0.3 percent above background and showed that the equipment would not introduce any spurious radioactivity to the sample measurements. A specimen of modern, coke-smelted cast iron with surface corrosion was then divided into four pieces and analyzed for carbon-14 activity, while being subjected to different cleaning methods. Theoretically, such samples should contain no carbon-14, except for undetermined amounts of modern carbon which may be added at the foundry and the contamination which may result from adsorbed atmospheric carbon dioxide and organic materials occluded in the corrosion layer. Sample Y–1503, which was not cleaned before milling, showed 1.2 $\pm$ 0.2 percent activity, relative to a modern standard; sample Y–1502, which was leached with acetic acid, had 1.3 $\pm$ 0.3 percent activity; sample Y–1501, which was dry peened and leached with acetic acid, had 0.7 $\pm$ 0.2 percent activity; Sample Y–1508, which was dry peened, acid leached, and baked at 200°C for thirty minutes under vacuum, had 1.2 $\pm$ 0.3 percent activity. These results are not entirely consistent, since Y–1501 gave the lowest reading and the best cleaned sample did not provide a better result than those to which fewer cleaning methods had been applied. It must therefore be assumed that adsorbed carbon dioxide is unimportant as a contaminating factor and that, for this specimen at least, no organic contamination of the corrosion layer had taken place. The observed (and variable) carbon-14 activity can be ascribed to contamination at the foundry, which would not be a factor when the iron is smelted with charcoal. In spite of the conclusion that contamination did not appear to play a part in the carbon-14 determinations of these iron samples, the removal of corrosion from all specimens was still considered to be necessary in view of the theoretical considerations described in chapter 4.

A second group of iron samples was analyzed to provide comparative historical and carbon-14 ages. Some thirty-five charcoal-smelted specimens, for which the date of manufacture was known with reasonable accuracy, were collected for this purpose from museums and interested individuals. Most of the samples were only a few centuries old (older specimens are difficult to obtain for experimental purposes), and some were too small for successful analysis. Nine samples were found to have sufficient carbon content and to provide a reasonable variation in ages; carbon-14 dates of these specimens were obtained. In eight cases, the results show close agreement between historically known ages and carbon-14 ages (see table 2). The results obtained for the three most recent samples (Y–1504, Y–1506, and Y–1509) provide carbon-14 ages which are on the average sixty years older than expected (taking into account the known variations of carbon-14 concentrations in tree ring material during this period). The tests described above are not sensitive to the possibility of adding C^{14}–free carbon to the sample, for instance from the alundum in the combustion boats. The

TABLE 2

CARBON-14 MEASUREMENTS OF IRON SAMPLES

LAB. NO.	SAMPLE DESCRIPTION	C-14 MEASUREMENT
Y-1500	Anthracite; no activity above background	0.0±0.3% activity
Y-1503	Modern, coke-smelted cast iron. Corrosion not removed	1.2±0.2 activity
Y-1502	Same as Y-1503. Leached with acetic acid.	1.3±0.3% activity
Y-1501	Same as Y-1503. Dry peened, leached with acetic acid	0.7±0.2% activity
Y-1508	Same as Y-1503. Dry peened, acid leached, and baked under vacuum	1.2±0.3 activity
Y-1504	Cast iron; Saugus, Mass. Manufactured A.D. 1648–78	350±60 B.P. A.D. 1600±60
Y-1506	Cast iron; Hopewell, Penna. Manufactured A.D. 1771–1845	200±60 B.P. A.D. 1750±60
Y-1509	Cast iron; Redding furnace, Penna. Manufactured A.D. 1761	180±60 B.P. A.D. 1770±60
Y-1510	Bloomery iron, Scotland. Manufactured A.D. 83–87	1850±80 B.P. A.D. 100±80

LAB. NO.	SAMPLE DESCRIPTION	C-14 MEASUREMENT
Y-1511	Cast iron; Sian, China. Han Dynasty (221 B.C.–A.D. 220)	2060±80 B.P. 110±80 B.C.
Y-1512	Cast iron; Hunan, China. Assumed to be 4th to 10th century A.D.	400±60 B.P. A.D. 1550±60
Y-1513	Cast iron; Honan, China. Warring States Period (480–221 B.C.)	2380±80 B.P. 430±80 B.C.
Y-1515	Cast iron; Szechwan, China. Han Dynasty (221 B.C.–A.D. 220)	2130±100 B.P. 180±100 B.C.
Y-1516	Carburized steel; Magdalenska gora, Yugoslavia. La Tène I (ca. 400–180 B.C.)	2130±60 B.P. 180±60 B.C.
Y-1505	Cast iron; Fort Kiowa?, S.D. A.D. 1870–1900	4.7±0.1% activity Apparent age ca. 25,000 B.P.
Y-1507	Cast iron; Fort Berthold, N.D. A.D. 1845–1885	43.6±0.5% activity Apparent age ca. 6700 B.P.

small difference of sixty years could be ascribed to this effect, but the difference is hardly outside the statistical error in the measurements of both iron and tree ring material. The contribution of old carbon, therefore, is neglibible. Sample Y–1512, an art object with a carbon–14 date at least 500 years younger than expected, can be assumed to have been misidentified on stylistic grounds; it is also possible that it is a deliberate copy of an earlier specimen, manufactured several centuries after the original. The consistency of the other results indicates that iron specimens can, indeed, be accurately dated by the carbon-14 method. The accuracy of dates in the 2000 year B.P. range, at which time there is no significant deviation of atmospheric carbon-14 activity from the predicted value, are especially convincing.

The final group of samples (Y–1505 and Y–1507) were chosen from that period in U.S. history when the iron industry was carrying out the conversion from charcoal to coke smelting. The observed carbon-14 activities—4.7 ± 0.1 percent (corresponding to an apparent age of *ca.* 25,000 B. P.) and 43.6 ± 0.5 percent, (apparent age *ca.* 6,700 B.P.)—are easily distinguishable from real carbon-14 dates. These specimens were either smelted with a charcoal-coke fuel mixture, or were manufactured by melting "coke iron" and "charcoal iron" together; their carbon-14 activity provides a measurement of the extent of this mixing. It is, of course, theoretically possible to obtain a spurious carbon-14 date for some charcoal-coke mixtures. Iron smelted in, say, 1870 with a mixture of 96 percent charcoal and 4 percent coke would have an apparent carbon-14 age of 400 years. On the basis of historical records, however, such examples can be assumed to have occurred infrequently, if at all, and most spurious dates can probably be detected on the basis of cultural association without much trouble.

A detailed account of the analysis of the samples follows. Details are included on the weights of samples in successive stages of preparation to provide information about sample loss; this information is summarized in appendix 2. Unless otherwise noted, all iron samples were cleaned as described above.

Sample Descriptions

Background Calibrations

Y–1500: *Anthracite* 0.0 ± 3% activity

Anthracite, analyzed to test the possible influence of the equipment on background radiation measurements.

Comment: The result shows no spurious radioactivity.

Y–1503: *Iron, Modern* 1.2 ± 0.2% activity

Modern, coke-smelted cast iron; part of a pot-bellied stove. Donated by M. D. Coe, Yale University.

Comment: the layer of surface corrosion was not removed, serving as a test for possible contamination. Comparison with Y–1502, Y–1501, and Y–1508 indicates that contamination is probably not due to the condition of the sample, but that modern carbon was introduced at the foundry.

Sample weight, corroded	181.3 g
Sample weight, milled	124.2 g
Amount of sample analyzed	74.0 g
Volume of extracted CO_2, purified, at S.T.P.	4.3 liters
Carbon content (calculated from yield)	3.1%

Y–1502: *Iron, Modern* 1.3 ± 0.3% activity

Modern coke-smelted cast iron; part of the same potbellied stove as Y–1503. (Donated by M. D. Coe, Yale University.)

Comment: The sample was leached with 25% acetic acid to test this method of cleaning. No improvement over Y–1503 was observed.

Sample weight, corroded	196.0 g
Sample weight, after leaching	187.3 g
Sample weight, milled	114.6 g
Amount of sample analyzed	74.5 g
Extracted CO_2, purified, at S.T.P.	3.7 liters
Approximate carbon content	2.6%

Y–1501: *Iron, Modern* 0.7 ± 0.2% activity

Modern, coke-smelted cast iron; part of the same potbellied stove as Y–1503. (Donated by M. D. Coe, Yale University.)

Comment: Corrosion was removed by dry peening and leaching with 25% acetic acid. The observed activity is inconsistent with Y–1503, Y–1502, and Y–1508 (although statistical considerations allow a small chance that the values are equal). It can therefore probably be assumed that variable contamination was introduced during manufacture.

Sample weight, corroded	138.5 g
Sample weight, cleaned	129.3 g

Sample weight, milled	89.5 g
Amount of sample analyzed	82.4 g
Extracted CO_2, purified, at S.T.P.	4.2 liters
Approximate carbon content	2.7%

Y–1508: *Iron, Modern* 1.2 ± 0.3% activity

Modern, coke-smelted cast iron; part of the same potbellied stove as Y–1503. (Donated by M. D. Coe, Yale University.)

Comment: Corrosion was removed by dry peening and leaching with 25% acetic acid; the milled sample was baked out under vacuum for 30 minutes to remove adsorbed carbon dioxide. No improvement over Y–1503 was observed.

Sample weight, corroded	345.2 g
Sample weight, cleaned	333.1 g
Sample weight, milled	208.2 g
Amount of sample analyzed	95.2 g
Extracted CO_2, purified, at S.T.P.	4.7 liters
Approximate carbon content	2.6%

Iron Samples of Known Age

Y–1504: *Saugus Ironworks* 350 ± 60 B.P.
A.D. 1600 ± 60

Cast iron; part of an adjustable crane hook from the Saugus Ironworks (42°30′N Lat, 71°30′W Long), near Hammersmith, Massachusetts.[1] Submitted by E. N. Hartley, Saugus Ironworks, and C. S. Smith, Massachusetts Institute of Technology. (Donated by the trustees of Saugus Ironworks.)

Comment: The foundry at Saugus started production in 1648 and closed down in 1678; the site has recently been reconstructed.

Sample weight, cleaned	380.0 g
Amount of sample milled	279.8 g
Sample weight, milled	223.5 g
Amount of sample analyzed	81.3 g
Extracted CO_2, purified, at S.T.P.	4.5 liters
Approximate carbon content	2.9%

1. Hartley 1957.

Y–1506: *Hopewell Furnace* 200 ± 60 B.P.
A.D. 1750 ± 60

Cast iron fragment from the debris around the furnace of the "iron plantation" at Hopewell, Pennsylvania (40°10′N Lat, 78°15′W Long). Submitted by P. W. Bishop, Smithsonian Institution, and Leland J. Abel, the excavator. (Donated by the Smithsonian Institution.)

Comment: Hopewell Furnace operated from 1771 to 1883;[2] sample Y–1506 was estimated to have been manufactured between 1771 and 1845.[3] Hopewell Village has been reconstructed as an historic site by the National Park Service.

Sample weight, cleaned	289.2 g
Sample weight, milled	211.3 g
Amount of sample analyzed	112.8 g
Extracted CO_2, purified, at S.T.P.	4.2 liters
Approximate carbon content	2.0%

Y–1509: *Redding Furnace* 180 ± 60 B.P.
A.D. 1770 ± 60

Cast iron stove plate, manufactured in the Redding Furnace (also known as the "Wicked Borrower"), in the Schuylkill-French Creek region, Pennsylvania. Submitted by Leonard G. Johnson, Bucks County Historical Society. (Donated by the trustees of the Society.)

Comment: The stove plate was manufactured in 1761, according to the date cast on the specimen.

Sample weight, cleaned	231.9 g
Sample weight, milled	162.3 g
Amount of sample analyzed	102.1 g
Extracted CO_2, purified, at S.T.P.	4.4 liters
Approximate carbon content	2.3%

2. Kurjack 1956.
3. Bishop 1964: personal communication.

Y–1510: *Roman fort, Scotland* 1850 ± 80 B.P.
A.D. 100 ± 80

Bloomery iron nails from a Roman Legionary fortress at Inchtuthil, Perthshire, Scotland (56°25′N Lat, 3°20′W Long). Submitted by H. Cleere, Iron and Steel Institute, London. (Collected and donated by the late Sir Ian Richmond, Oxford University.)

Comment: The fort at Inchtuthil formed part of a system of Scottish Highland forts designed by Agricola. Built in A.D. 83, the fort was dismantled in A.D. 87, following the withdrawal of Roman troops from Scotland. The nails,[4] part of a seven-ton cache, were buried to prevent their falling into the hands of Scots; as a result, a sizable portion of them were only lightly oxidized.

Sample weight, cleaned	1232.1 g
Sample weight, milled	579.3 g
Amount of sample analyzed	579.3 g
Extracted CO_2, purified, at S.T.P.	2.4 liters
Approximate carbon content	0.2%

Y–1511: *Han Dynasty, Sian, China* 2060 ± 80 B.P.
110 ± 80 B.C.

Cast iron fragment from the grating of a stove from Sian, China (34°20′N Lat, 109°E Long). Chicago Natural History Museum catalogue number 120988. Submitted by Kenneth Starr, Chicago Natural History Museum. (Donated by the trustees of the museum.)

Comment: The stove was assumed to have been manufactured during the Han Dynasty (221 B.C.–A.D. 220); the carbon-14 date corroborates this identification.

Sample weight, cleaned	187.4 g
Sample weight, milled	151.6 g
Amount of sample analyzed	102.1 g
Extracted CO_2, purified, at S.T.P.	6.1 liters
Approximate carbon content	3.2%

4. Angus, Brown, and Cleere 1962.

Y–1512: *Hunan Province, China* 400 ± 60 B.P.
A.D. 1550 ± 60

Cast iron from the base plate of a Chinese statuette from Hunan Province (*ca.* 28°N Lat, 112°E Long). (Anonymous donor.)

Comment: This art object, which had been dated as being from the fourth to the tenth century A.D., apparently on stylistic grounds, has been exhibited in several museums. The carbon-14 date indicates that it was assigned to the wrong stylistic period, or that it was a deliberate Ming Dynasty copy of an earlier original. The possibility also exists that the base plate was added at a later date, and that the carbon-14 date does not refer to the statue itself.

Sample weight, cleaned	200.4 g
Sample weight, milled	172.8 g
Amount of sample used	97.8 g
Extracted CO_2, purified, at S.T.P.	3.5 liters
Approximate carbon content	1.9%

Y–1513: *Warring States Period, Honan, China*
2380 ± 80 B.P.
430 ± 80 B.C.

Cast iron fragment from a tomb group near Loyang, Honan Province, China (34°48′N Lat, 112°24′E Long). Collected in the 1930's (Bishop White collection of the Royal Ontario Museum). Submitted by the late William Todd, Royal Ontario Museum, and C. S. Smith, Massachusetts Institute of Technology. (Donated by the Royal Ontario Museum.)

Comment: The iron fragments were curved in cross-section and were apparently cast to reinforce round poles. The tomb group contained a large number of bronze fittings and was assumed to date from the time of the feudal state of Han, Warring States Period (480–221 B.C.); the carbon-14 date corroborates this identification.

Sample weight, cleaned	46.4 g
Sample weight, crushed	46.0 g
Amount of sample analyzed	42.3 g
Extracted CO_2, purified, at S.T.P.	2.5 liters
Approximate carbon content	3.1%

Y–1515: *Han Dynasty, Szechwan, China* 2130 ± 100 B.P.
180 ± 100 B.C.

Cast iron fragments from a group of cist tombs at Chia-shan-Chai, Li-fan Hsien, Szechwan Province, China (31°45′N Lat, 103°45′E Long). Collected in the 1940's during archaeological excavations by the Academia Sinica. Submitted by Kwang-chih Chang, Yale University. (Donated by S. S. Ling, National Taiwan University.)

Comment: The iron fragments formed part of a collection which includes pottery assumed to be of the Han Dynasty (221 B.C.–A.D. 220); the carbon-14 date corroborates this identification.

Sample weight, cleaned	32.7 g
Sample weight, crushed	32.4 g
Amount of sample analyzed	32.4 g
Extracted CO_2, purified, at S.T.P.	1.6 liters
Approximate carbon content	2.6%

Y–1516: *La Tène I, Yugoslavia* 2130 ± 60 B.P.
180 ± 60 B.C.

Fragments of a carburized steel sword and scabbard from Magdalenska gora, Yugoslavia (46°5′N Lat, 14°30′E Long), Tumulus V, graves 19–20. Collected between 1905 and 1914 by the Duchess Paul Friedrich of Mecklenburg-Schwerin. Submitted by Hugh M. Hencken, Harvard University. (Donated by the Peabody Museum of Harvard University.)

Comment: The sample had been almost completely converted to oxide and consisted of a large number of small, friable pieces, encrusted with clayey dirt which had become cemented by the iron oxide. The sample was boiled and repeatedly washed in distilled water and then broken into quarter-inch fragments for analysis. Upon cleaning, a composite rivet (?) of gold and copper was found in the debris. The carbon-14 date corroborates the previously assumed cultural age, La Tène I (*ca.* 400–180 B.C. in Western Europe), which had been assigned to the specimen.

Sample weight, cleaned and crushed	216.0 g
Amount of sample analyzed	216.0 g
Extracted CO_2, purified, at S.T.P.	2.7 liters
Approximate carbon content	0.66%

Nineteenth Century Iron Samples

Y–1505: *Fort Kiowa (?), S. D.* 4.7 ± 0.1% activity
Apparent age *ca.* 25,000 B.P.

Cast iron stove part from Fort Kiowa (?), Lyman County, South Dakota (44°N Lat, 99°20′W Long). Excavated for the Missouri River Basin Survey (catalogue number 39LM57–4). Submitted by Clifford Evans, Smithsonian Institution. (Donated by the Smithsonian Institution.)

Comment: The identification of this site by the excavator is questionable; it is now thought to have been a late trading post, operating between 1870 and 1900 A.D., and not Fort Kiowa itself.[5] The observed carbon-14 activity indicates that the specimen was manufactured by melting together "coke iron" and "charcoal iron" (by, for example, adding a small amount of "charcoal iron" scrap to a batch of "coke iron"), or by fueling the furnace with a coke-charcoal mixture in a ratio of about 95:5.

Sample weight, cleaned	198.0 g
Sample weight, milled	139.2 g
Amount of sample analyzed	97.5 g
Extracted CO_2, purified, at S.T.P.	5.5 liters
Approximate carbon content	3%

5. Evans 1964: personal communication.

Y–1507: *Fort Berthold, N. D.* 43.6 ± 5% activity
Apparent age *ca.* 6700 B.P.

Cast iron wagon thimble from the second Fort Berthold, McLean County, North Dakota (47°30′N Lat, 101°30′W Long). Excavated for the Missouri River Basin Survey (catalogue number 32LM2–1254; USNM number 430526). Submitted by Clifford Evans, Smithsonian Institution. (Donated by the Smithsonian Institution.)

Comment: The site of the second Fort Berthold was occupied from 1845 to 1885. The observed carbon-14 activity leads to the same conclusion as for Y–1505, but with a "coke iron"—"charcoal iron"—"charcoal iron," or a coke-charcoal ratio of about 55:45.

Sample weight, cleaned	245.6 g
Sample weight, milled	184.9 g
Amount of sample analyzed	83.3 g
Extracted CO_2, purified, at S.T.P.	4.2 liters
Approximate carbon content	2.7%

7

CARBON-14 DATING OF SLAG

THE ARCHAEOLOGIST who studies Iron Age sites often finds only stains in the soil where iron artifacts should have been; this is especially true of sites in the acid soils of, for example, savannah regions. If iron was manufactured at a site, large quantities of slag are often present, while only a few iron specimens may be recovered. Slag, which consists primarily of silica, is almost indestructible in comparison with iron. In addition, about four times as much slag as iron is produced in smelting furnaces; this slag generally remains concentrated in large dumps on the manufacturing site, while the iron is distributed among the inhabitants of a wider region.

Certain kinds of slag contain significant amounts of carbon. This is especially the case with cinder, which is a mixture of slag and unburnt charcoal formed in the bottom of bloomery furnaces. Datable cinder can be identified by inspection, since the charcoal nodules are often clearly visible. The charcoal can be extracted mechanically in the field or in the laboratory; forceps and a thin knife blade or scalpel is all that is required to remove a sample for carbon-14 dating. This procedure provides the archaeologist with an additional method for dating Iron Age sites.

The successful dating of iron slag has been reported by

the Stockholm Radioactive Dating Laboratory[1] and by Geochron Laboratories;[2] a similar procedure was followed in a field study of Iron Age sites in South Africa.[3] In addition to the dating of charcoal samples from the excavations, iron slag samples were dated; it was also found possible to date slag from copper smelting sites. Copper furnaces produce cinder which is almost indistinguishable from the variety produced in the smelting of iron, although the copper slag forms above the metal product, not below it.

The South African field study was carried out in the district of Palabora, in the Northeastern Transvaal. The area today is the scene of a fast-growing industrial mining complex, concentrated around extensive iron and copper ore deposits in a hill known as Loolekop—the site of an extinct volcano. Scattered around Loolekop in an area roughly thirty miles in diameter are numerous secondary volcanic hills, most of which have euphonious Sotho names—Kgopolwe, Matsepe, Molotho, Sealene, Tsankare—and the sides of which are stepped with the stone-walled living terraces of abandoned Iron Age settlements. Loolekop itself shows no sign of previous habitation, but the numerous mineshafts, galleries and adits which have been uncovered in the course of recent mining activities lead mining engineers to estimate that the Iron Age peoples removed some 10,000 tons of rock from the hill. Low, grass-covered mounds of slag, scattered in abundance in the plain around the living sites, attest to an extensive iron and copper smelting industry. Carbon-14 dates from a number of sites (table 3) indicate that Iron Age miners were active in the area as early as A.D. 800. It is historically known that the BaPalabora, a Sotho-speaking tribe of Iron Age people, were living in this area until 1904, when they were resettled in the nearby Makushane reservation. Preliminary analysis and comparison of the pottery from a tenth century habitation site at

1. Engstrand and Östlund 1962; Östlund and Engstrand 1963.
2. Krueger and Weeks 1965.
3. Stuiver and van der Merwe 1968.

Kgopolwe (Y–1637, Y–1638, Y–1639, and Y–1662), a late nineteenth century site at the same hill, and the modern pottery which is infrequently produced in Makushane, show closely related styles of manufacture and decoration. It can be assumed, then, that the ancestors of the BaPalabora, or peoples with a similar cultural tradition, have occupied the Palabora area for more than 1,000 years.

The problem of studying this 1,000–year cultural tradition, in the light of the imperfect archaeological knowledge of the area, is complicated by the close similarity in pottery styles through the entire sequence. Further excavation and detailed study of the pottery will serve to alleviate this problem, enabling archaeologists to estimate the age of habitation sites on the basis of pottery types. Studying the development of metallurgical techniques, however, is more difficult. The Palabora district is littered with smelting sites which may be anywhere from fifty to 1,000 years old or more, while the surface finds of slag can serve only to determine whether iron or copper were smelted at a particular location. Sampling, therefore, is a major problem. It was found that the dating of slag can be extremely useful as a survey tool in distinguishing between manufacturing sites of different ages. Slag deposits, even when totally covered with soil, were easily spotted by the rugged varieties of grasses which grow on them. Sufficient cinder for a carbon-14 measurement can be collected from such deposits in a matter of hours, where days would be required to excavate the site. Since the slag is clearly associated wth past human activity (which would not be the case with charcoal collected in the same manner), such carbon-14 dates can be associated with the smelting operations at the site. In this manner it was established that most of the readily visible smelting sites in the area date from the past 100 years (Y–1660 and Y–1661); in future intensive studies of the earlier occupations, these sites (and those similar to them in appearance) can be avoided.

In the course of excavating an iron smelting site (SPK–IV at Kgopolwe, Palabora), slag was collected for dating along with charcoal from the furnace. A date of A.D. 1430 ± 60 (Y–1658) was obtained from the slag; a date of A.D. 1670 ± 60 (Y–1657) from the charcoal. The difference in the dates may not be as large as it appears, in view of the wide variation in carbon-14 activity during the past 400 years. However, the slag deposit was estimated to contain about fifty tons of material, equivalent to about 25,000 pounds of manufactured iron. If operations at this site did, in fact, continue for two centuries, this would mean an annual yield of 125 pounds of smelted iron—enough to keep a small community supplied with agricultural implements and weapons. The stratigraphic excavation and dating of slag heaps would be the only reliable method which could provide further information on the extent and duration of smelting operations at this and other sites in the area. Coupled with studies of the habitation sites, valuable information could be gained about the rate of metal consumption by the population.

A detailed account of the carbon-14 dates obtained for charcoal and slag samples from the Iron Age deposits at Palabora follows; this information is summarized in table 3.

Palabora Iron Age Series

Seven samples of charcoal, two samples of iron slag, and one sample of copper slag were collected from various Iron Age sites near the town of Palabora, Transvaal, South Africa (23°55′S Lat, 31°10′E Long). Collected in 1965 by N. J. van der Merwe and M. Stuiver.[4]

Y–1635: *Loolekop Mine-Shaft-1* 950 ± 60 B.P.
A.D. 1000 ± 60

Charcoal from a mine-shaft, 70 feet deep, at Loolekop, Palabora.

Comment: The shaft, a narrow trench about fifty feet in length at the top and seventy feet deep in the middle, was completely revealed in cross-section as a

4. *Ibid.*

TABLE 3

CARBON-14 DATES FOR IRON AGE SITES AT PALABORA, TRANSVAAL

LAB. NO.	SAMPLE TYPE AND LOCATION	CARBON-14 AGE
Y-1635	Charcoal from 70-foot mineshaft at Loolekop, Palabora	950±60 B.P. A.D. 1000±60
Y-1636	Charcoal from 20-foot mineshaft at Loolekop, Palabora	1180±80 B.P. A.D. 770±80
Y-1637	Charcoal from hut floor of early occupation at site SPK-III, Kgopolwe, Palabora	910±60 B.P. A.D. 1040±60
Y-1638	Charcoal from hut floor of early occupation at site SPK-III	990±80 B.P. A.D. 960±80
Y-1639	Charcoal found immediately below hut of later occupation at site SPK-III	850±60 B.P. A.D. 1100±60
Y-1662	Charcoal from hut floor of later occupation at site SPK-III	820±80 B.P. A.D. 1130±80
Y-1658	Charcoal from floor of iron bloomery furnace at site SPK-IV, Kgopolwe, Palabora	280±60 B.P. A.D. 1670±60
Y-1657	Slag from 50-ton slag heap surrounding furnace at site SPK-IV	520±60 B.P. A.D. 1430±60
Y-1660	Slag from small deposit around iron bloomery furnace, still standing on surface at Matsepe, Palabora	80±60 B.P. A.D. 1870±60
Y-1661	Slag from small deposit around copper furnace, still standing on surface, at Molotho, Palabora	60±120 B.P. A.D. 1890±120 (counting time 200 mins.)

result of blasting in the course of modern mining activities. The sample was collected from the deepest part of the shaft within hours after the blasting took place. The carbon-14 age is in good agreement with the ages of Y–1636, Y–1637, Y–1638, Y–1639, and Y–1662; these measurements constitute a series which includes the oldest known dates for the Iron Age in South Africa.

Y–1636: *Loolekop Mine-Shaft-2* 1180 ± 80 B.P.
A.D. 770 ± 80

Charcoal from a gallery at the bottom of a mine-shaft, twenty feet deep, at Loolekop, Palabora.

Comment: The charcoal came from an extensive charcoal deposit (probably the result of fire-setting) on the floor of a horizontal gallery, some twenty feet in length, at the bottom of the shaft; collecting was accomplished in the same manner as for Y–1635. This is the oldest known date for the Iron Age in South Africa.

Y–1637: *Kgopolwe III, Lower Floor-1* 910 ± B.P.
A.D. 1040 ± 60

Charcoal from surface of clay floor of earlier occupation at Kgopolwe, Palabora, Site SPK-III (SPK-III/A:21″–24″ below datum).

Comment: The clay floor, part of a hut and living area, was stratified below the floor of a later occupation (Y–1639 and Y–1662). The date is consistent with Y–1638, from the same floor, and with dates for the upper floor.

Y–1638: *Kgopolwe III, Lower Floor-2* 990 ± 80 B.P.
A.D. 960 ± 80

Charcoal from the surface of a clay floor of the earlier occupation at Kgopolwe, Palabora, Site SPK-III (SPK-III/C: 25″–27″ below datum).

Comment: The sample was taken from the same floor as Y–1637, from a different part of the site. Consistent with Y–1637, Y–1639, and Y–1662.

Y–1639: *Kgopolwe III, Upper Floor-1* 850 ± 60 B.P.
A.D. 1100 ± 60

Charcoal from immediately below a clay hut floor of the later occupation at Kgopolwe, Palabora, Site SPK-III (SPK-III/D: 21½″ below datum).

Comment: The clay floor beneath which this sample occurred is of the same type as for Y–1637 and Y–1638, and was stratified about six inches above it. Consistent with Y–1637, Y–1638, and Y–1662.

Y–1662: *Kgopolwe III, Upper Floor-2* 820 ± 80 B.P.
A.D. 1130 ± 80

Charcoal from surface of clay hut floor of later occupation at Kgopolwe, Palabora, Site SPK-III (SPK-III/CD: 18″ below datum).

Comment: The charcoal was collected from the surface of the floor, beneath the collapsed walls of the hut, and dates the end of the later occupation at SPK-III. Consistent with Y–1637, Y–1638, and Y–1639.

Y–1658: *Kgopolwe IV, Iron Furnace-1* 280 ± 60 B.P.
A.D. 1670 ± 60

Charcoal from the floor of an iron bloomery furnace at Kgopolwe, Palabora, Site SPK-IV.

Comment: The sample represents the last time iron was smelted in the furnace.

Y–1657: *Kgopolwe IV, Iron Furnace-2* 520 ± 60 B.P.
A.D. 1430 ± 60

Charcoal in iron slag, from slag heap surrounding furnace at Kgopolwe, Palabora, Site SPK-IV.

Comment: The slag sample, taken from a deposit of some fifty tons, represents a date during the period of smelting operations at the site. Consistent with Y–1658.

Y–1660: *Matsepe Iron Furnace* 80 ± 60 B.P.
A.D. 1870 ± 60

Charcoal in iron slag, from a small deposit around an iron bloomery furnace at Matsepe, Palabora.

Comment: The furnace was still standing on the surface, one of a number of similar sites.

Y–1661: *Molotho Copper Furnace* 60 ± 120 B.P.
A.D. 1890 ± 120

Charcoal in copper slag, from a small deposit around a copper smelting furnace at Molotho, Palabora.

Comment: The furnace was still standing on the surface, as in the case of the furnace at Matsepe. The sample was counted for 200 minutes only.

8

CONCLUSIONS

THE RESULTS of the research reported in this book are twofold. One phase of the research concerns problems in laboratory technique and includes the successful development of a method for the carbon-14 dating of iron artifacts, a further demonstration of the application of carbon-14 measurements to the dating of iron slag, and the application of the latter technique to the dating of copper slag. The other phase of the research concerns archaeological problems and is designed to draw attention to the potential of these dating methods—either through the dating of specimens of iron or slag to provide solutions to specific problems in Iron Age chronologies, or by pointing out the many possibilities where these techniques can be fruitfully applied.

The laboratory investigation (chapters 4 through 7) provides a body of results which confirms the theoretical expectation that specimens of all the alloys of iron can be dated through the extraction of carbon which is contained in the metal. The carbon-14 dates of a series of specimens show close agreement between measured and known ages, while experiments with modern specimens, in varying states of corrosion, demonstrate that the carbon-14 dating of iron is nearly free of the influence of contaminants. The utility of dating iron slag by the carbon-14 method, which has previously been reported by other researchers, is further demonstrated by the dating of a number of samples of this type, while the same procedure is used for the first time for the dating of copper slag.

From the archaeological point of view, an introduction to the historical events and metallurgical innovations of the Iron Age is provided (chapters 2 and 3). By combining available knowledge of the prehistory and history of iron-using cultures with information about specific metallurgical techniques employed at various times and places, a picture of the Iron Age emerges which differs substantially from prevalent archaeological sources. Special attention is paid to the substantial difference between the traditions of iron-working in the West (the direct process) and the East (the indirect process), the internal evolution of each of these traditions, and the historical implications of the relationship between the two traditions.

The double-pronged approach of the research is united in the solution of chronological problems of the Iron Age through the carbon-14 dating of a number of iron and slag samples which relate to such problems. The majority of the dated iron samples came from assemblages which had been collected before the development of the carbon-14 dating method; their dating serves to provide the first absolute age measurements for the archaeological sequences of which they form a part. In other cases, the potential solutions of similar problems is demonstrated.

The Laboratory Research

Problems and Results

The results of the laboratory research demonstrate that iron specimens can be accurately dated by the carbon-14 method. Close correlations between known ages and carbon-14 ages, especially in the 2,000–year age range, not only confirm the theoretical expectations that iron can be dated, but also show that the measurements are not affected to any detectable extent by contaminants (chapter 6). This conclusion is supported by the carbon-14 activity measurements obtained in a controlled experiment with corroded modern iron (Y–1501, Y–1502, Y–1503, and Y–1508). It is also significant that many of the dated samples (e.g., Y–1511, Y–1513, Y–1515, and Y–1516) were not only corroded, but had been gathering dust and radioactive fallout on museum shelves for more than twenty years. These contaminants are easily removed from iron specimens and do not constitute a problem in the dating of the specimens; charcoal, if stored under similar conditions, would have been unsuitable for carbon-14 analysis.

Carbon-14 measurements of coke-smelted iron (Y–1505 and Y–1507) demonstrate that the use of fossil fuels can be readily detected and that the ratio of fossil fuel to charcoal can be determined wherever such mixtures were used.

As a further addition to age determinations, the carbon-14 analysis of iron samples also provides measurements of the carbon contents of the specimens (90% estimated accuracy). Few of the dated iron specimens had previously been analyzed by chemical or spectographic means; the laboratory results served to identify the alloy type of at least one specimen (Y–1516: La Tène, carburized steel), and to confirm the presumed alloy types of several others (Y–1511, Y–1513, and Y–1515: Chinese Chou and Han dynasties, cast iron). These samples date from crucial periods in the history of iron metallurgy, but relatively few metallurgical analyses of similar specimens have been performed. The combination of an absolute age measurement and an alloy identification provides significant information in the reconstruction of the metallurgy of these periods.

Iron slag, which has previously been dated by a number of investigators, was used in the present project as a material for dating in an archaeological investigation of Iron Age chronologies in South Africa (Y–1657 and Y–1660). The type of slag known as cinder contains quantities of occluded carbon which derives from the smelting fuel; this carbon can be extracted mechanically and treated in the same manner as charcoal samples. The same technique was also used for the first time in the carbon-14 dating of copper slag (Y–1661). The dates obtained from slag samples were consistent with a series of dates for charcoal samples (Y–1635–39, Y–1662, and Y–1689) from the same sequence.

Potential Uses

From the point of view of laboratory analysis, the dating of iron as well as iron and copper slag has great potential. The carbon-14 dating of materials like wood and charcoal has by now become a byword in archaeology; the dating of iron and slag adds a new dimension to the scope of the existing techniques. The analysis of slag samples differs only slightly from conventional carbon-14 work, and its impact concerns the archaeologist more than the laboratory scientist; the carbon-14 dating of iron, however, demands further sophistication in laboratory analysis and considerably expands the inventory of datable specimen types. The kinds of problems which can be fruitfully investigated as a result of the development of iron and slag dating include the following:

1. *Museum collections.* The dating of a few iron specimens from archaeological assemblages can serve to establish the age of important collections. Since contamination from dust and corrosion does not constitute a problem, it does not matter how long the specimens have been in storage.
2. *Archaeological excavations.* Both iron and slag specimens can be dated as companion techniques to conventional carbon-14 or other dating methods.
3. *Collector's items.* Wherever sufficient material can be removed from an iron art object without spoiling its esthetic qualities, its age can be firmly established.

4. *"Coke iron."* The transition from charcoal smelting to coke (or, in some instances, coal) smelting in various areas can be investigated by means of carbon-14 measurements of iron specimens. Charcoal-coke mixing ratios can be determined, while the time of the transition can be established if some independent means of dating the object is available.
5. *Alloy types.* These can be identified as a result of the extraction of carbon from iron specimens in the course of carbon-14 analysis. Combined with chemical or metallographic analyses, wherever necessary, valuable metallurgical information can be obtained.

Carbon-14 analyses of this kind can go far towards providing more extensive chronological information for the study of the Iron Age. It would be appropriate at this point to interject a note on the cost of analysis of iron and slag samples. Carbon from slag is treated in exactly the same manner in the laboratory as conventional carbon-14 samples; no extra cost is therefore involved. The analysis of iron specimens, on the other hand, as performed during this investigation, exceeded by about 60 percent the cost of analyzing charcoal samples. The additional cost could, however, be reduced to about 10 percent by a laboratory which undertakes the routine analysis of iron specimens. With the addition of a series of efficient dust traps, the carbon extraction equipment described in chapter 5 can be attached to the gas-purification equipment of a carbon-14 laboratory, while the suitability of iron samples can be determined with a carbon analyzer. The use of equipment of this kind would reduce the cost of analyzing iron samples to the level of conventional charcoal samples, with the sole additional expense of cleaning and milling the specimen.

The dating of iron is not currently being carried out on a routine basis by carbon-14 laboratories. At the time of writing, however, the Conservation-Analytical Laboratory of the Smithsonian Institution and the Historical Committee of the Verein Deutscher Eisenhüttenleute have reported interest in carrying on with this type of research.

Prehistory and History of the Iron Age: Problems and Results

The Two Metallurgical Traditions

The survey of metallurgical techniques involved in the production of iron alloys, combined with current knowledge of the prehistory and history of the Iron Age (chapters 2 and 3), provide a perspective against which the potential of the dating methods described in this volume can be viewed.

Two basic metallurgical traditions are involved in the Iron Age—the direct process of the West and the indirect of the East. In the West, the development of the basic techniques of the direct process signaled the beginning of the Iron Age. Achieving its initial impact around 1500–1400 B.C., this process involves the production of bloomery iron by the bloomery process, and the associated technique for the manufacture of steel from bloomery iron by the cementation process. From its center of origin in the Hittite empire, the direct process spread through large parts of Europe, Africa, and Asia by the beginning of the Christian era. In China, on the other hand, a radically different metallurgical tradition, known as the indirect process, was developed during the sixth century B.C. This process involves the production of cast iron in blast furnaces, and the manufacture of wrought iron and steel from cast iron by decarburization. The Chinese tradition of iron metallurgy remained confined to the East until the late fourteenth century, when the indirect process was adopted in Europe, and the massive iron industry of today was initiated.

Viewing the archaeological sequences of the Iron Age in terms of two separate metallurgical traditions, each with an internal series of developments, has important implications in prehistoric and historic studies. Archaeologists have generally paid inadequate attention to the lessons which can be learnt from metallurgy, thus allowing the perpetuation of metallurgical fallacies and erroneous reconstructions of Iron Age events. It is, among other problems, not sufficient to discuss the diffusion of iron-working knowledge without specifying the technological level of such knowledge. It has been pointed out, from the relatively meager metallurgical information available, that the bloomery process and cementation process did not diffuse simultaneously to all areas (chapter 3). Similarly, techniques like quenching and tempering did not achieve immediate distribution to all iron-using cultures extant at the time of their discovery. The highly important difference between the direct and indirect process has also been inadequately recognized. As a result of these inadequacies, a number of crucial questions have gone substantially unasked or unanswered on the part of prehistorians and historians. Perhaps the most important of these involves the appearance of iron production in China. Did the Sarmatians, or some other people, acquaint the Chinese with the uses of iron, or was the radically different tradition a totally independent invention? Similarly, did the direct process develop independently in thirteenth and fourteenth century Europe, or did the Chinese tradition merely undergo a period of adaptation in the West? What metallurgical techniques were used by Iron Age cultures in Africa? These questions, and many more, will have to be answered before adequate reconstructions of Iron Age events can be achieved.

Laboratory Results

The results obtained during the laboratory study help to answer the kinds of questions posed above, or help to illustrate how their solutions can be achieved. Most of the laboratory results, especially those obtained in the course of developing the techniques of dating iron by the carbon-14 method, were necessarily of a corroborative nature; a number of results which contribute to the solution of prehistoric and historic problems were, however, obtained.

Carbon-14 dates of 430 ± 80 B.C. (Y–1513), 180 ± 100 B.C. (Y–1515), and 110 ± 80 B.C. (Y–1511) were obtained for iron samples of the Chinese Chou and Han Dynasties (chapter 6). These specimens had been collected during the 1930's and 1940's, before the development of carbon-14 dating, and their presumed ages (which were corroborated) had been arrived at on the basis of cultural associations. As a result of the current political situation, the area from which they were collected is no longer accessible for study to Western scholars. The carbon-14 dating of iron specimens constitutes the only available means of obtaining absolute age measurements for these and similar collections. In addition, the measurements of carbon contents (3.1%, 2.6%, and 3.2%, respectively) which resulted from the carbon extraction procedure, confirm the opinions of the collectors that the specimens were made of cast iron, and further demonstrate the antiquity of cast iron in China. In a European context, a carbon-14 date of 180 ± 60 B.C. (Y–1516) was obtained for iron of the La Tène I period in Yugoslavia, which had been excavated in the early 1900's. The carbon-14 date corroborates the opinion of the excavator regarding the position of the collection in the chronological sequence; the carbon content of the specimen (0.66%) identifies it as carburized steel, which confirms previous information about the metallurgical skills of La Tène peoples. Further corroborative carbon-14 dates were obtained for iron specimens from relatively well-dated collections. These include a date of A.D. 100 ± 80 (Y–1510) for bloomery iron nails (carbon content, 0.2%) from a Roman fort in Scotland, previously dated to A.D. 87; a date of A.D. 1600 ± 60 for cast iron (2.9% C) from Saugus, Massachusetts, previously dated between A.D. 1648 and 1678; a date of A.D. 1750 ± 60 (Y–1506) for cast iron (2.0% C) from Hopewell, Pennsylvania, previously dated between A.D. 1771 and 1845; a date of A.D. 1770 ± 60 (Y–1509) for cast iron (2.3% C) from the Redding furnace, Pennsylvania, previously dated to A.D. 1761. In another case, a date of A.D. 1550 ± 60 (Y–1512) for a Chinese cast iron statuette (1.9% C), cast doubt on the authenticity of the specimen, which had previously been assumed to date from the fourth to tenth century A.D.

Carbon-14 activity measurement of nineteenth century iron specimens from U.S. frontier forts yielded results of 4.7 ± 0.1 percent (Y–1505) and 43.6 ± 0.5 percent (Y–1507). These measurements indicate that the specimens in question were smelted with fuel mixtures of coke and charcoal, or that they were manufactured by melting together "coke iron" and "charcoal iron." In either case, the carbon-14 measurements fix the coke–charcoal mixing ratios at 95:5 and 55:45, respectively. These measurements illustrate how the use of coke (or coal) smelting can be identified and measured.

The carbon-14 dating of iron and copper slag served as a companion technique to the dating of charcoal in establishing a chronology for the Iron Age sites at Palabora, in the Transvaal, South Africa (chapter 7). Dates for charcoal samples established ages of *circa* A.D. 800 to 1100 for a series of sites, representing the earliest known Iron Age occupation in South Africa. Although the sequence is not complete, dates of A.D. 1430 ± 60 (Y–1657) and A.D. 1870 ± 60 (Y–1660) for iron slag samples, and a date of A.D. 1890 ± 120 (Y–1661) for a copper slag sample indicate that the prehistoric Iron Age occupation of the area may have been continuous for 1,000 years. In addition, an age difference of some 200 years between the date for charcoal from the furnace at a smelting site (Y–1658), and the date for iron slag from the surrounding deposit (Y–1657), allowed for a calculation of the approximate annual yield of iron production at the site.

The advantages of dating iron and slag specimens are of primary importance to the archaeologist. Unlike charcoal, such materials are clearly the result of human activity, and thus eliminate many of the problems of demonstrating associations between carbon-14 dates and cultural events.

Archaeological Potential

The library and laboratory study of prehistoric and historic problems of the Iron Age reported here has significant potential for archaeological studies.

It is hoped that the treatment of Iron Age events in terms of the development of metallurgical processes will produce a different orientation toward this period of man's history. While such an orientation is by no means new, it has received inadequate treatment on the part of archaeologists. Concentrated attention should be paid to the techniques of iron-working used by prehistoric and historic peoples; it is not enough to specify only whether a given culture made use of iron or not.

The carbon-14 dating of iron can provide the archaeologist with a valuable new research tool. Not only does it expand the scope of the carbon-14 dating method for future investigations of the Iron Age, but it allows the archaeologist to investigate collections for which datable materials had not been available until now. Carbon-14 analyses can also serve as an investigative method in studying the use of fossil fuels in iron production. In addition, the carbon-14 dating of iron necessarily draws attention to the specific type of iron alloy under investigation. Some knowledge of the alloy type is required to determine the suitability of an iron specimen for dating, while a carbon-14 analysis also provides a measurement of the carbon content of the specimen. If further information is obtained by means of chemical, spectographic, and metallographic analysis, it will bring the investigator closer to an accurate reconstruction of Iron Age events.

The dating of iron slag provides another valuable research tool. Slag can be dated to establish chronological sequences; it can also be used as a surveying device, since a minimum of excavation of a slag deposit is required to obtain an absolute age for a smelting site. In addition, extensive excavation and dating of slag deposits may enable the investigator to calculate production yields of smelting operations. The dating of copper slag has the same potential as the analysis of iron slag, while expanding the time range which can be investigated to the beginning of the Chalcolithic.

A number of problems, which can be solved by means of these techniques, immediately come to mind. It would be possible, for example, to test the authenticity of cast iron statuettes which are claimed to be of the Roman period (see chapter 3), as well as to verify the results of chemical analyses which indicate that the Romans occasionally used coal to smelt iron. More carbon-14 dates for early Chinese iron specimens in archaeological collections would add significantly to our incomplete knowledge of the initial events in the history of the direct process. If possible, specimens from the Sarmatian-Chinese contact area should be analyzed to throw more light on the question of whether iron-working was an independent invention in China. In North America, the authenticity of iron claimed to be of Viking origin could be tested; in Africa, the dating of iron and slag could fill the large gaps in the available information on the Iron Age of the continent.

The above list is obviously only an indication of the kinds of problems that can be solved by the dating of iron and slag. The individual investigator would no doubt be able to think of many more problems in his own area of interest. The solution of such problems by means of the techniques which have been described could go far in providing greater accuracy in the reconstruction of prehistoric and historic events during the Iron Age—the most crucial period of modern man's technological development.

APPENDIX 1

Equipment Used for Carbon Extraction

ITEM	SOURCE
Electric furnace, Hoskins FH 305	Hoskins Manufacturing Co. Detroit, Michigan
Transformer, for use with furnace	Sorgel Electric Co. Milwaukee, Wisconsin
Rheostat, for use with furnace	Ward Leonard Electric Co. Mt. Vernon, New York
Combustion tubes, zircon	Laboratory Equipment Corp. St. Joseph, Michigan
Combustion tube baffles, porcelain	McDanel Refractory Porcelain Co. Beaver Falls, Pennsylvania
Combustion boats, quartz, custom made	Thermal American Fused Quartz Co. Montville, New Jersey

Suggested Additional Equipment

Carbon analyzer, Leco 598–500	Laboratory Equipment Corp. St. Joseph, Michigan

APPENDIX 2

IRON SAMPLE WEIGHT

SAMPLE NO.	SAMPLE WEIGHT, CLEANED, (GRAMS)	SAMPLE WEIGHT, MILLED, (GRAMS)
Y-1501	129.3	89.6
Y-1502	187.3	114.6
Y-1503	181.3	124.2
Y-1504	279.8	223.5
Y-1505	198.0	139.2
Y-1506	289.0	211.3
Y-1507	228.5	184.9
Y-1508	333.1	208.2
Y-1509	231.9	162.3
Y-1510	1232.1	579.3
Y-1511	187.4	151.6
Y-1512	200.4	172.8
Y-1513	46.4	crushed
Y-1515	32.7	crushed
Y-1516	216.0	crushed

O$_2$ YIELDS, AND CARBON CONTENTS

AMOUNT OF SAMPLE ANALYZED (GRAMS)	CO_2 YIELD, UNPURIFIED (LITERS)	CO_2 YIELD, PURIFIED (LITERS)	CARBON CONTENT CALCULATED FROM CO_2 YIELD %
82.4	5.7	4.2	2.7
74.5	4.7	3.7	2.6
74.0	4.7	4.3	3.1
81.3	5.4	4.5	2.9
97.5	5.9 (4.7) *	4.4	3.0
112.8	5.1	4.2	2.0
83.3	4.5	4.2	2.7
95.2	6.0	4.7	2.6
102.1	5.8 (4.5) *	3.4	2.3
579.3	2.8	2.4	0.2
102.1	7.0 (4.0) *	3.5	3.2
97.8	4.5	3.5	1.9
42.3	3.4	2.5	3.1
32.4	1.7	1.6	2.6
216.0	3.2	2.7	0.66

* Amount of CO_2 was reduced to the figure in parentheses before purification.

BIBLIOGRAPHY

Agrawal, D. P.; Kusumgar, S.; and Lal, D.
1956 Tata Institute radiocarbon date list III. *Radiocarbon* 7:291–95.

Agricola, Georgius
1556 *De re metallica.* Original published in Basle. Translated by H. C. and L. H. Hoover. 1st ed., London: Mining Magazine, 1912; 2d ed., New York: Dover, 1950.

Aitchison, Leslie
1960 *A history of metals.* 2 vols. London: Macdonald and Evans.

Aitken, M. J.
1961 *Physics and archaeology.* New York and London: Interscience Publishers.

Alexander, W., and Street, A.
1956 *Metals in the service of man.* London: Pelican.

Angus, N. S.; Brown, G. T.; and Cleere, H. F.
1962 The iron nails from the Roman legionary fortress at Inchtuthil, Perthshire. *Journal of the Iron and Steel Institute* 200:956–68.

Anstee, J. W., and Biek, L.
1956 Pattern-welded iron swords. *Nature* 178:1432–33.

Ashton, T. S.
1924 *Iron and steel in the Industrial Revolution.* Manchester: Manchester University Press.

Baker, H. G.
1949 Blast furnace construction and costs in 1740. *Newcomen Society, Transactions* 24:113–20.

Barendsen, G. W.
1955 *Ouderdomsbepaling met radioactieve koolstof.* The Hague: Uitgeverij Excelsior.
Bellamy, C. V.
1904 A West African smelting house. *Journal of the Iron and Steel Institute* 56:99–126.
Biringuccio, Vannoccio
1540 *Pirotechnia.* Translated by C. S. Smith and M. T. Gnudi. New York: Basic Books, 1959.
Boerschmann, E.
1931 *Chinesische Pagoden.* Berlin and Leipzig: de Gruyter.
Brown, B., and Herbert, A.
1950 Iron objects from Azarbaijan. *Man* 50:79.
Butts, Allison
1963 Metallurgy. In *Encyclopaedia Britannica* 15:306–18.
Carpenter, H. C. H., and Robertson, J. M.
1930 The metallography of some ancient Egyptian implements. *Journal of the Iron and Steel Institute* 121:417–54.
Chang, Kwang-chih
1963 *The archaeology of ancient China.* New Haven, Conn. Yale University Press.
Chao Ch'üan-chia
1944 Lu-shan Han-tai Yeh-t'ieh-ch'ang yih-chih Chih fa-hsien yü K'ao-cheng [Excavation and study of Han Dynasty iron workshop at Lu-shan]. *Shih-Hsüeh Tsa Chih* 1:57.
Childe, V. G.
1944 Archaeological ages as technological stages. *Journal of the Royal Anthropological Institute* 74:1–19.
Clark, Grahame
1961 *World prehistory.* Cambridge: Cambridge University Press.
Clark, J. G. D.
1952 *Prehistoric Europe: The economic basis.* London: Methuen.
Clark, J. D.
1962 The spread of food production in Sub-Saharan Africa. *Journal of African History* 3:211–28.
Cline, Walter
1937 *Mining and metallurgy in Negro Africa.* Menasha, Wis.: George Banta.
Coghlan, H. H.
1941 Prehistoric iron prior to the dispersal of the Hittite empire. *Man* 41:74–80.
1956 *Notes on prehistoric and early iron in the Old World.* Pitt Rivers Museum Occasional Papers on Technology, no. 8. Oxford: Pitt Rivers Museum.

Crawhall, T. C.
1933 Iron working in the Sudan. *Man* 33: 41–43.
Creswell, K. A. C.
1956 *Bibliography of arms and armour in Islam.* London: Royal Asiatic Society.
Damon, P. E.; Long, A.; and Grey, D. C.
1966 Fluctuation of atmospheric C^{14} during the last six millenia. *Journal of Geophysical Research* 71: 1055–63.
Davidson, B.
1959 *The lost cities of Africa.* Boston and Toronto: Little, Brown and Co.
Day, St. John V.
1877 *The prehistoric use of iron and steel.* London: Trübner.
de Boer, J. H.
1953 *The dynamical character of adsorption.* Oxford: The Clarendon Press.
Desch, C. H.
1928 Reports on the metallurgical examination of specimens for the Sumerian Committee of the British Association. *Reports of the British Association for the Advancement of Science 1928–39.* London: British Assoc. for Advancement of Science.
1956 Metallurgical aids to archaeology. *Newcomen Society, Transactions* 27: 121–30.
de Vries, H.
1955 Purification of CO_2 for use in a proportional counter for C^{14} age measurements. *Applied Scientific Research* B5: 387–400.
1958 Variation in concentration of radiocarbon with time and location on earth. *Koninklijke Nederlandse Akademie van Wetenschappen, Verhandelingen* B61: 94.
de Vries, H., and Barendsen, G. W.
1953 A new technique for radiocarbon dating by a proportional counter filled with carbon dioxide. *Physica* 19: 987–1003.
Dickmann, H.
1956 *Zweitausend Jahre Eisenguss, und die Entwicklung des Kupolofens.* Düsseldorf: Bibliothek des Vereins Deutscher Eisenhüttenleute.
Dono, T.
1937 The chemical investigation of the ancient metallic culture of the Orient. *Journal of the Faculty of Science, Imperial University of Tokyo* Section I, 3: 287.
Dyck, W.
1967 Recent developments in radiocarbon dating. *Current Anthropology* 8: 349–51.

Elsasser, W. E.; Ney, P.; and Winckler, J. R.

1956 Cosmic-ray intensity and geomagnetism. *Nature* 178:1226–27.

Engstrand, L. G., and Östlund, H. G.

1962 Stockholm natural radiocarbon measurements IV. *Radiocarbon* 4:115–36.

Fagan, Brian M.

1961 Pre-European ironworking in Central Africa. *Journal of African History* 2:199–210.

1963*a* The Iron Age sequence in the southern province of Northern Rhodesia. *Journal of African History* 4:157–77.

1963*b* The Kalomo/Chomo Iron Age project. *South African Archaeological Bulletin* 18:3–19.

1964 The Greefswald sequence: Bambandyanalo and Mapungubwe. *Journal of African History* 5:337–61.

1965 *Southern Africa in the Iron Age.* New York: Praeger.

Fagg, B.

1961 An ancient site in Niger Province. *Bulletin of News, Historical Society of Nigeria* 5, no. 4:3.

Fieldhouse, W. J.; May, T.; and Wellstood, F. C.

1931 *A Romano-British industrial settlement near Tiddington, Stratford-upon-Avon.* Birmingham: W. I. Rodway.

Fisher, Douglas A.

1963 *The epic of steel.* New York: Harper & Row.

Forbes, R. J.

1942 *Bibliographia antiqua, philosophia naturalis, vol. 2: metallurgy.* Leiden: Nederlandsch Instituut voor het Nabije Oosten.

1950*a* *Man the maker.* New York: Henry Schuman.

1950*b* *Metallurgy in antiquity.* Leiden: E. J. Brill.

1952 *Bibliographia antiqua, philosophia naturalis,* supplement 1. Leiden: Nederlandsch Instituut voor het Nabije Oosten.

1954 "Extracting, smelting, and alloying." In *A history of technology,* ed. Charles Singer *et al,* vol. 1, pp. 572–99. Oxford: The Clarendon Press.

1956 "Metallurgy." In *A history of technology,* ed. Charles Singer *et al,* vol. 2, pp. 41–80. Oxford: The Clarendon Press.

Frankfort, H.

1950 Early iron in Iraq. *Man* 50:100.

Friend, J. A. N.

1926 *Iron in antiquity.* London: Griffin and Co.

Gabel, Creighton

1965 African prehistory. In *Biennial review of Anthropology,* ed. Bernard J. Siegel, pp. 40–83. Stanford, Calif.: Stanford University Press.

Gilles, J. W.
1958 Versuchssmelze in einem vorgeschichtlichen Rennofen. *Stahl und Eisen* 79:1690–95.

Gloag, J., and Bridgewater, D.
1948 *A history of cast iron in architecture.* London: Allen and Unwin.

Goel, P. S., and Kohman, T. P.
1962 Cosmogenic carbon-14 in meteorites and terrestrial ages of "finds" and craters. *Science* 136:875–76.

Goldsmith, J. N., and Hulme, E. W.
1948 Sketch of the history of natural draught furnaces. *Newcomen Society, Transactions* 23:1–12.

Gowland, W.
1912 The metals in antiquity. *Journal of the Royal Anthropological Institute* 42:235–87.

Greenberg, J. H.
1963 *The languages of Africa. International Journal of American Linguistics* 29, no. 1, part 2.

Hannack, G.
1930 Japanische Damaststahl. *Beiträge zur Geschichte der Technik und Industrie* 20:87–90.

Haring, A.; de Vries, A. E.; and de Vries, H.
1958 Radiocarbon dating up to 70,000 years by isotopic enrichment. *Science* 128:472–73.

Hartley, E. N.
1957 *Ironworks on the Saugus.* Norman, Okla.: University of Oklahoma Press.

Hiernaux, J., and Maquet, E.
1960 Cultures préhistoriques de l'age des metaux au Ruanda-Urundi et au Kivu, Congo Belge. *Académie Royale des Sciences d'Outre-Mer* 60:1–102.

Hodges, Henry
1964 *Artifacts.* London: John Baker.

Honan Bureau of Culture Archaeological Team
1965 The casting process of the Han Dynasty iron ploushshares as indicated by the pottery moulds and cores unearthed from the remains of Wan Ch'eng at Nanyang, Honan Province. *Wen Wu* 7:1–12.

Hough, W.
1926 *Fire as an agent in human culture.* United States National Museum Bulletin 139. Washington: U. S. National Museum.

Hughes, E. E., and Mann, W. B.
1964 The half-life of carbon-14: Comments on the mass-spectroscopic method. *International Journal of Applied Radiation and Isotopes* 15:97–100.

Hulme, E. W.
1939 Prehistoric and primitive iron smelting, part I: From meteoric iron to 1000 B.C. *Newcomen Society* 18:181–92.
1943 Prehistoric and primitive iron smelting, part II: The crucible processes of the East. *Newcomen Society, Transactions* 21:23–30.
Inami, Hakusui
1948 *Nippon-Tō: The Japanese sword.* Tokyo: Cosmo.
Ingstad, Helge
1964 Vinland ruins prove Vikings found the New World. *National Geographic* 126:708–34.
Inskeep, R. R.
1962 Some Iron Age sites in Northern Rhodesia. *South African Archaeological Bulletin* 17:136–80.
Jenkins, Rhys
1924 Notes on the early history of steel making in England. *Newcomen Society, Transactions* 3:16–40.
1935 The reverberatory furnace with coal fuel. *Newcomen Sosiety, Transactions* 14:67–82.
Johannsen, O.
1925 *Geschichte des Eisens.* Düsseldorf: Verlag Stahleisen. (2d edition 1955.)
1933 Gab es in Karolingerzeit schon Hochofen? *Stahl und Eisen* 53:1039–40.
Karlén, I.; Olsson, I. U.; Kallberg, P.; and Killicci, S.
1964 Absolute determination of the activity of two C^{14} dating standards. *Arkiv Geofysik* 4:465–71.
Kigoshi, Kunihiko, and Hasegawa, Hiroichi
1965 Secular variation of atmospheric radiocarbon concentration and its dependence on geomagnetism. *Journal of Geophysical Research* 71:1065–72.
Kohmann, T. P., and Goel, P. S.
1963 Terrestrial ages of meteorites from cosmogenic C^{14}. In *Radioactive Dating.* Proceedings of the symposium on radioactive dating, Athens, 1962. Vienna: International Atomic Energy Agency.
Koşay, H. Z.
1944 *Ausgrabungen von Alaca Hüyük.* Ankara.
1951 *Les Fouilles d'Alaca Hüyük, 1937–39.* Ankara.
Kořan, Jan
1965 Survey of the evolution of the indirect process of ironmaking in the Czech lands. *Acta historiae rerum naturalium nec non technicarum* 1, special issue: 37–63. Prague: Czechoslovak studies in the history of science.

Krueger, Harold W., and Weeks, C. F.
1965 Geochron Laboratories, Inc., radiocarbon measurements I. *Radiocarbon* 7:47–53.

Kulp, J. L.
1952 The carbon-14 method of age determination. *Scientific Monthly* 75:259–67.

Kurjack, Dennis C.
1956 *Hopewell Village.* National Park Service Historical Handbook Series, no. 8. Washington, D. C.: National Park Service.

Larsen, H., and Rainey, F.
1948 *Ipiutak and Arctic whale hunting culture.* Anthropological papers of the American Museum of Natural History 42.

Levin, M. G., and Sergeyev, D. A.
1964 The penetration of iron into the Arctic. In *The archaeology and geomorphology of Northern Asia,* ed. H. N. Michael ed., pp. 319–26. Arctic Institute of North America, Anthropology of the North: Translation from Russian Sources 5. Toronto: University of Toronto Press.

Libby, Willard F.
1952 *Radiocarbon dating.* Chicago: University of Chicago Press.
1955 *Radiocarbon dating.* 2d edition. Chicago: University of Chicago Press.

MacKay, C.; Pandow, M.; and Wolfgang, R.
1963 On the chemistry of natural radiocarbon. *Journal of Geophysical Research* 68:3929–31.

Mallowan, M. E. L.
1936 Excavations at Tell Chagar Bazar. *Iraq* 3, part 1.
1937 Excavations at Tell Chagar Bazar. *Iraq* 4.

Mann, W. B.; Marlow, W. F.; and Hughes, E. E.
1961 The half-life of carbon-14. *International Journal of Applied Radiation and Isotopes* 11:57–67.

Maryon, H., and Plenderleith, H. J.
1954 Fine metal-work. In *A history of technology,* ed. Charles Singer *et al.* vol 1, pp. 623–62. Oxford: The Clarendon Press.

Mason, Revil J.
1965 The origin of South African society. *South African Journal of Science* 61:255–67.

Mason, Revil J., and van der Merwe, N. J.
1964 Radiocarbon dating of Iron Age sites in the Southern Transvaal: Melville Koppies and Uitkomst Cave. *South African Journal of Science* 60:142.

May, T.
1904 *Warrington's Roman remains.* Warrington.

Needham, Joseph
1958 *The development of iron and steel technology in China.* London: The Newcomen Society.

Neuberger, Albert
1930 *The technical arts and sciences of the ancients.* Trans. H. L. Brose. New York: Macmillan.

Olsson, I. U.; Karlén, I.; Turnbull, A. H.; and Prosser, N. J. D.
1962 A determination of the half-life of C^{14} with a proportional counter. *Arkiv Fysik* 22: 237–55.

Olsson, I. U.; and Karlén, I.
1963 The half-life of C^{14} and the problems which are encountered in absolute measurements on beta-decaying gases. In *Radioactive dating.* Proceedings of the symposium on radioactive dating, Athens 1962. Vienna: International Atomic Energy Agency.

Östlund, H. G., and Engstrand, L. G.
1963 Stockholm natural radiocarbon measurements V. *Radiocarbon* 5:203–27.

Pandow, M.; MacKay, C.; and Wolfgang, R.
1960 The reaction of atomic carbon with oxygen: Significance for the natural radio-carbon cycle. *Journal of Inorganic Nuclear Chemistry* 14:153–58.

Parr, J. Gordon
1958 *Man, metals and modern magic.* New York: Collier.

Partington, J. R.
1935 *Origins and development of applied chemistry.* London: Longmans Green.

Peake, H.
1933 The origin and early spread of ironworking. *Geographical Review* 23:639–52.

Percy, John
1864 *Metallurgy: Iron and steel.* London: John Murray.

Perry, S. H.
1944 *The metallography of meteoric iron.* United States National Museum Bulletin 184. Washington, D.C.: U.S. National Museum.

Phillips, E. D.
1965 *The royal hordes: Nomad peoples of the steppes.* London: Thames and Hudson.

Pleiner, R.
1965 Iron smelting experiments in protohistoric furnaces. *XI Congres International D'histoire Des Sciences,* Sommaires section No. 4 at 5:551–3. Warsaw: Warszawska Drukarnia Naukowa.

Plenderleith, H. J.
1956 *The conservation of antiquities and works of art.* London: Oxford University Press.
Posnansky, M.
1961 Iron Age in East and Central Africa—Points of comparison. *South African Archaeological Bulletin* 16:134–36.
Rainey, F., and Ralph, E.
1959 Radiocarbon dating in the Arctic. *American Antiquity* 24:365–74.
Ralph, E. R., and Stuckenrath, R., Jr.
1960 Carbon-14 measurements of known age samples. *Nature* 188:185–87.
Read, T. T.
1934*a* The early casting of iron. *Geographical Review* 24:544–54.
1934*b* Metallurgical fallacies in archaeological literature. *American Journal of Archaeology* 38:382–89.
1936 Early Chinese metallurgy. *The Metal Industry* 48:308.
Richardson, H. C.
1934 Iron, prehistoric and ancient. *American Journal of Archaeology* 38:555–83.
Rickard, T. A.
1932 *Man and metals.* New York and London: McGraw Hill.
Ritson, J. A. S.
1958 Metal and coal mining. In *A History of technology,* ed. Charles Singer *et al.,* vol. 4, pp. 64–98. Oxford: The Clarendon Press.
Robinson, K. R.
1959 *Khami Ruins.* Cambridge: Cambridge University Press.
1961 An early Iron Age site from the Chibi District, Southern Rhodesia. *South African Archaeological Bulletin* 16:75–102.
1963 Further excavations in the Iron Age deposits at the Tunnel Site, Gokomere Hill, Southern Rhodesia. *South African Archaeological Bulletin* 18:155–71.
Schmitz, Von Fritz
1930 Orientalischer Damaststahl. *Beiträge zur Geschichte der Technik und Industrie* 20:81–86.
Schubert, H. R.
1956 Early refining of pig-iron in England. *Newcomen Society. Transactions* 28:59–76.
1957 *History of the British iron and steel industry.* London: Routledge & Kegan Paul.
1958 "Extraction and production of metals: Iron and steel." In *A history of technology,* ed. Charles Singer *et al.,* vol. 4, pp. 99–117. Oxford: The Clarendon Press.

Singer, Charles, *et al.*, ed.
1954–58 *A history of technology.* 4 vols. Oxford: The Clarendon Press.
Smith, Cyril S.
1960 *A history of metallography.* Chicago: University of Chicago Press.
Smith, Cyril S., and Forbes, R. J.
1957 Metallurgy and assaying. In *A history of technology,* ed. Charles Singer *et al.,* vol. 3, pp. 27–71. Oxford: The Clarendon Press.
Smythe, J. A.
1938 Ancient Sussex iron blooms—a metallographic examination. *Newcomen Society, Transactions* 17:197–203.
Straker, E.
1931 *Wealden iron.* London: G. Bell and Sons.
Stuiver, M.
1961 "Variations in radiocarbon concentration and sunspot activity." *Journal of Geophysical Research* 66:273–6.
1965 Carbon-14 content of 18th–and 19th century wood: Variations correlated with sunspot activity. *Science* 149:533–35.
Stuiver, M., and van der Merwe, N. J.
1968 Radiocarbon chronology of the Iron Age in Sub-Saharan Africa. *Current Anthropology* 9:54–58.
Stuiver, M., and Suess, H. E.
1966 On the relationship between radiocarbon dates and true sample ages. *Radiocarbon* 8:534–40.
Suess, H. E.
1965 Secular variations of the cosmic-ray-produced carbon-14 in the atmosphere and their interpretations. *Journal of Geophysical Research* 70:5937–52.
Suess, H. E. and Wänke, H.
1962 Radiocarbon content and terrestrial age of twelve stony meteorites and one iron meteorite. *Geochimica et Cosmochimica, Acta* 26:475–80.
Summers, Roger
1958 *Inyanga: prehistoric settlements in Southern Rhodesia.* Cambridge: Cambridge University Press.
1961 The Southern Rhodesian Iron Age. *Journal of African History* 2:1–13.
1963 *Zimbabwe, a Rhodesian mystery.* London: Nelson.
Summers, Roger; Robinson, K. R.; and Whitty, A.
1961 *Zimbabwe excavations 1958.* Occasional Papers of the National Museum of Southern Rhodesia 23A.

Turekian, K. K.
1961 The role of trace-element geochemistry in archeology. *Yale Scientific Magazine* 35, no. 4: 6–11.

Tylecote, R. F.
1962 *Metallurgy in archaeology.* London: Edward Arnold.

van der Merwe, N. J.
1965 Carbon-14 dating of iron: A new archaeological tool. *Current Anthropology* 6: 475.

van der Merwe, N. J., and Stuiver, M.
1968 Dating iron by the carbon-14 method. *Current Anthropology* 9: 48–53.

Wainwright, G. A.
1936 The coming of iron. *Antiquity* 10: 5–24.
1945 Iron in the Napatan and Meroitic ages. *Sudan Notes and Records* 26: 5–36.

Wang Yü-Ch'üan
1957 Iron farm tools in ancient China. *China Reconstructs* 6, no. 2: 10–11.

Watt, D. E.; Ramsden, D.; and Wilson, H. W.
1961 The half-life of carbon-14. *International Journal of Applied Radiation and Isotopes* 11: 68–74.

Welcher, F. J., ed.
1963 *Standard methods of chemical analysis.* 6th ed. Princeton: D. van Nostrand.

Wheeler, Sir Mortimer
1959 *Early India and Pakistan.* London: Thames and Hudson.
1966 *Civilizations of the Indus Valley and beyond.* London: Thames and Hudson.

Willis, E. H.; Tauber, H.; and Münnich, K. O.
1960 Variations in the atmospheric radiocarbon concentration over the past 1300 years. *Radiocarbon* 2: 1–4.

Willis, E. H.
1963 Radiocarbon dating. In *Science in archaeology,* (D. Brothwell and E. Higgs, editors). New York: Basic Books.

Wynne, E. J., and Tylecote, R. F.
1958 An experimental investigation into primitive iron smelting technique. *Journal of the Iron and Steel Institute* 190: 339–48.

Zimmer, G. F.
1916 The use of meteoric iron by primitive man. *Journal of the Iron and Steel Institute* 94: 306–49.

INDEX